Global Forest Visualization

This book project examines global forest monitoring as a means to understand the promises and problems of global visualization for climate management.

Specifically, the book focuses on Global Forest Watch, the most developed and widely available forest-monitoring platform, created in 1997 by the World Resource Institute. Forest maps are always political as they visualize power relations and form the grid within which forests become commodities. This dislocation of the idea of the forest from its literal roots in the ground has generated problems for forest visualization efforts designed to empower local communities. This book takes a critical humanistic approach to this problem, combining methods from the fields of rhetoric and media studies to suggest solutions to these problems for designers and users of platforms like the Global Forest Watch. To explain why global views of forests can be disempowering, the book relies on biopolitical and rhetorical theories of panopticism and how these views unfold a different violence on different regions of the Earth in relation to colonial history. Using this theoretical framework, the book explains the historical process by which forests came to be classified, quantified, and mapped on a global scale. Interviews with end-users of global forest visualization platforms reveal if and how these platforms support local action. Lastly, the book provides rhetorical solutions to articulate global and local views of forests without reducing one view to the other. These solutions involve looking to forests themselves for clues about how to generate more broadly effective and resilient visualizations.

This book will be of great interest to students and scholars of forest studies, climate change, science communication, visualization studies, environmental communication, and environmental conservation.

Lynda Olman is Professor of English at the University of Nevada, Reno, USA. She is the author of *Scientists as Prophets* (2013) and the editor of *Global Rhetorics of Science* (2023), as well as other books on the rhetoric of science. Her current work focuses on improving risk visualizations to support robust decision-making on environmental and climatic issues.

Birgit Schneider is Professor of Knowledge Cultures and Media Environments at the Potsdam University Institute for Arts and Media, Germany. Her current research concentrates on the visual communication of climate since 1800 and a genealogy of climate change visualization between science, aesthetics, and politics.

Routledge Focus on Environment and Sustainability

Agricultural Digitization and Zhongyong Philosophy
Creating a Sustainable Circular Economy
Yiyan Chen, Hooi Hooi Lean, and Ye Li

EU Trade-Related Measures against Illegal Fishing
Policy Diffusion and Effectiveness in Thailand and Australia
Edited by Alin Kadfak, Kate Barclay, and Andrew M. Song

Food Cultures and Geographical Indications in Norway
Atle Wehn Hegnes

Sustainability and the Philosophy of Science
Jeffry L. Ramsey

Food Cooperatives in Turkey
Building Alternative Food Networks
Özlem Öz and Zühre Aksoy

The Economics of Estuary Restoration in South Africa
Douglas J. Crookes

Urban Resilience and Climate Change in the MENA Region
Nuha Eltinay and Charles Egbu

Global Forest Visualization
From Green Marbles to Storyworlds
Lynda Olman and Birgit Schneider

For more information about this series, please visit: www.routledge.com/Routledge-Focus-on-Environment-and-Sustainability/book-series/RFES

Global Forest Visualization

From Green Marbles to Storyworlds

Lynda Olman and Birgit Schneider

First published 2024
by Routledge
4 Park Square, Milton Park, Abingdon, Oxon OX14 4RN

and by Routledge
605 Third Avenue, New York, NY 10158

Routledge is an imprint of the Taylor & Francis Group, an informa business

British Library Cataloguing-in-Publication Data
A catalogue record for this book is available from the British Library

ISBN: 978-1-032-45400-9 (hbk)
ISBN: 978-1-032-45401-6 (pbk)
ISBN: 978-1-003-37677-4 (ebk)

DOI: 10.4324/9781003376774

Typeset in Times New Roman
by Apex CoVantage, LLC

Contents

Acknowledgments

The authors wish to thank the following people:

Birgit Müller for sharing the anecdotes from her work with farmers in Nicaragua that provided the spark that fired this project. The Rachel Carson Center at Ludwig-Maximilians-Universität in Münich for funding the joint fellowship in the fall of 2017 that resulted in the proposal for this book, and the Alexander von Humboldt Foundation for the generous Friedrich Wilhelm Bessel Award that enabled us to research and write it together in Berlin for the first six months of 2022. The staff of the reading room at the Staatsbibliothek in Berlin for pulling a fabulous assortment of German forest maps from their collection that inspired our investigation into the history of European forest commodification. The staff of Global Forest Watch for their support and insights on many layers. All our interviewees, the power users of Global Forest Watch. Lola Pfeiffer for consulting and informing us with the extremely useful bibliography from her MA thesis, "Cartographies of the Unmappable" (Potsdam University, 2021). Gabriele Dürbeck, Philip Hüpkes, and Simon Probst for editing and publishing an early version of our core argument in *Narratives of Scale in the Anthropocene*, and for pointing us toward ways of telling stories that will hopefully get us past the "endless cyclopean war story from above" (Haraway). Jana Belmann and Julie Hagedorn for their assistance in transcribing interviews and proofreading, and Jonathan Gray and Myriel Milicevic for discussing forest issues with us.

All research activities involving human subjects that are reported in this book were carried out with the approval of the University of Nevada, Reno, Institutional Review Board (exemption 1809994–2) and the University of Potsdam Ethics Commission (Nr. 71/2021).

1 The Promises and Problems of Global Forest Visualization

A Nicaraguan farmer goes out to his field only to find a European man standing there with a GPS unit. The European is there to audit the REDD+ (Reducing Emissions from Deforestation and Degradation) project that his company is funding via a third-party non-profit organization that provides carbon offsets via reforestation projects. The auditor felt no need to contact the farmer or ask for permission to be on his land; he merely used the non-profit's smartphone app to navigate his rental car to this field. He speaks such poor Spanish that he cannot explain to the farmer on whose land he is trespassing what he is doing there or why. Notwithstanding, he will go back and report to his company that the trees the farmer ripped out to plant a nourishing corn crop are doing poorly—a report that will result in the farmer getting paid so much less per "carbon credit" than projected that he will have to rip the plantation out and replant corn next season to feed his family.[1]

When we heard this anecdote from a colleague of ours doing anthropological research in Nicaragua, we were fascinated and troubled. At the time, we were working together on a fellowship at the Rachel Carson Center (RCC) in Munich on global forest visualization, following up on work we had done independently on technical visualizations of climate change. As scholars in the rhetoric of science (Olman) and media ecology (Schneider), we were each concerned by the problem we had observed over and over again of sophisticated visual arguments about climate change failing to result in any substantial political action. Just before we went to the RCC, Birgit had stumbled across Global Forest Watch (GFW), the best developed and most widely used open-source forest-monitoring platform. GFW was created in 1997 by the World Resources Institute in response to calls at the 1996 Conference of Parties (COP-2) for better cooperation among government, non-profit, and corporate sectors to fight climate change. Advances in satellite imaging enabled GFW to quickly compose global images of forests and to provide rapid feedback on changes in the global canopy—both deforestation due to fires and legal or illegal logging and reforestation sponsored by national and transnational agencies. According to GFW, these deforestation maps led to real political action to protect tropical rainforests, crucial to mitigating climate change,

DOI: 10.4324/9781003376774-1

in the study sites the organization had worked with most closely: Indonesia, Cameroon, Gabon, and Canada. After learning all of this, we proposed a project to the RCC to study how GFW had managed to accomplish what so many visualizations of climate change had failed to achieve in the past.

So, our colleague's anecdote about the Nicaraguan farmer and the European monitor over coffee in the RCC lounge rocked us back in our chairs a bit. We had gotten excited about the power of the green marble—that is, the world's forests visualized as an organic, living whole, the 'lung of the planet'[2]—to effect political change in the ways that blue marble images of the Earth from space had galvanized environmental movements in the 1970s. However, we hadn't yet looked at the dark side of the green marble: it hadn't dawned on us that the ability to visualize a forest on the other side of the world at a computer terminal conferred at least some power on the people sitting at that terminal to interfere with that forest in the real world—irrespective of the interests of the people and other beings living in that forest on a day-to-day basis, all in the name of saving the planet.

Accordingly, we decided to narrow the focus of our project to this problem. We framed two key research questions within the problem because obviously it is incredibly, perhaps inexhaustibly, large and complex—touching on issues in forestry and climate services, history, economics, international aid, geography, ecology, anthropology, soil science, atmospheric physics, etc. We couldn't possibly address all of those dynamics in one book. What we *could* do with our backgrounds and expertise was focus on the problem of visualization and ask the following research questions:

1. What aspects of interactive green marble platforms like GFW were helping import transnational geopolitics into local forests, particularly in the Global South?
2. If designers and users of those platforms did not like those politics but wanted to keep referencing the green marble in their work, were there tactics they could use to articulate it with local views of their forests of concern (FOC), without either one canceling the other out?

The Geopolitics of Global Visualizations of Forests

In answering these questions, we are responding to work in the fields of environmental studies, critical geography, social anthropology, and science communication that calls for the 'downscaling' of global images of climate change to galvanize effective action at local levels.[3] We are also responding to calls within forest studies for mixed-methods approaches to understanding deforestation threats at a multiplicity of scales.[4] Finally, we are engaging literature on Indigenous data sovereignty that has raised concerns about the exportation of local information about territory, people, and culture up to the transnational/global level.[5] The key contribution we can make to this conversation, at least here at the outset of our project, is that zooming between global and local views of forests is neither as seamless nor as innocent an act of visualization as it first appears.

We knew before we began our project that forest maps are always political, as they are always made by someone for some purpose—usually for taking inventory of timber and other forest products for trade. This pattern goes back to the very first forest maps made in Europe in the Middle Ages by princes tallying up wealth and taxes, and it goes right on through to green investors calculating the carbon stocks buried in the boreal forests in Canada. We address this history in Chapter Two, and then we go on to explain why these global maps of forests can be disempowering to the people who live in them, relying on the political philosophies of Michel Foucault, Donna Haraway, and others.[6]

The situation gets more complex, however, when we add an interactive zoom tool to global forest maps, that is, when we can transition in a matter of seconds from a green marble view of the planet to a patch of deforestation in Nicaragua no larger than 30 m^2. How do we connect the disparate and distorted views at those different scales into a smooth, cinematic experience? Here the works of humanities scholars such as Tim Ingold and Zachary Horton, as well as science philosopher Bruno Latour, help us make sense of the optical illusions generated by the zoom tool in global forest visualization platforms.[7] This critical background leads us to the following conundrum: How do you reconcile global and local views of a particular forest if they are incommensurable, say, if the global view puts trees where locals recognize none, or vice versa?[8] Historically, the way this conundrum has been solved in climate visualization is to erase or reduce the local view in favor of the global, a solution that presents problems for climate justice because the people who live in local climate-change hotspots are predominantly non-white and economically vulnerable.[9] These are precisely the visualization problems this book seeks solutions for.

Alternatives to the Green Marble

Because of these problems, some environmental justice (EJ) activists have recommended rejecting the green marble, and the satellite surveillance that comes along with it, altogether. In its place, they have framed a couple of alternative ways of viewing forests, which we discuss in Chapters Two, Three, and Four under the heading of *cosmograms*. A cosmogram, according to science studies scholar John Tresch, is an image of the world that encapsulates a particular cosmology—a way of understanding the world and how to live in it. As an example, the medieval European cosmogram below shows both how the artist believed the universe was constructed, as concentric rings with the Earth at the center, and how we should live in it—subservient to a creator god whose angel servants are literally turning the cranks of creation (Figure 1.1). The

Figure 1.1 A medieval cosmogram attributed to the atelier of the Catalan Master of St Mark, from the Yates Thompson manuscript at the British Library.

Source: Figure adapted from Wikimedia Commons.

cosmogram also gives the viewer important clues about the community that made it and believed it to be beautiful and valuable—in the choice of colors for skin and clothing, the decorative elements, and the pigments themselves.

The green marble is a cosmogram, meaning that it is an image that encapsulates a whole worldview, a way of understanding the world's forests and valuing them—in this case, as commodities to be traded or banked against climate change. Understanding this principle, EJ activists have proposed alternative cosmograms to the green marble that tell different stories about forests: primarily counter-maps and dwellings, examples of which we provide in Chapter Three. These alternative maps disrupt the global view by scribbling over it, blowing out the scale so it can't be comprehended in a glance, coding it, or rendering it non-visually, as a 'soundscape' or perhaps a haptic map like the one in Figure 1.2. This navigation map is only intelligible if employed in situ in a particular archipelago in the Pacific Ocean, at which point it charted

Figure 1.2 A Micronesian navigation map made of wood, sennit fiber, and cowrie shells. From the collection of the Phoebe A. Hearst Museum of Anthropology at the University of California, Berkeley.

Source: Figure adapted from Wikimedia Commons.

for the mariner the locations of islands (cowrie shells) and the currents that ran among them (sticks); it could be used by touch at night and optically in the day.

What This Book Adds to the Conversation

Since these solutions have already been discovered, what's the point of this book? The REDD+ projects that seemed so promising 15 years ago at COP-13 are now being replaced with grassroots forest management efforts due to problems like the ones we noted at the start of this chapter.[10] Meanwhile, GFW has increased support for alternative, on-the-ground methods of monitoring deforestation (cf. interviews in Chapters Five and Six). And environmental scholars are increasingly promoting localized, 'nature-based' solutions to climate change and climate justice, even as talk of fixing global warming by spraying metric tons of calcium carbonate into the atmosphere continues.[11] What then can we contribute at this juncture? This: we discovered in our discussions with the GFW staff and power users of the platform that many EJ activists find value in global maps of their FOC for various reasons which we will detail in Chapter Four, many of which have to do with making Indigenous communities and their concerns visible. No scholarship, to our knowledge, has yet wrestled with the conundrum presented here: namely, how local communities—especially Indigenous communities in the Global South—can integrate green marble-type images of their FOC with their own local forest cosmograms without reducing one to the other or deleting one. So, that is what we set out to do in this book. We determined that the best way to do it was simply to ask the users of GFW how *they* were doing it. Accordingly, in Chapter Four we describe the results of our case study with designers and users of GFW. We interviewed three developers/managers from GFW and four users from sites around the world: Cameroon, Indonesia, Peru, and Georgia. Additionally, we conducted think-aloud protocols with three of the users to observe how they operated the GFW platform, with special attention to their zooming activities. From this research, we collected not only confirmations of known problems with green marble-type platforms but also a series of ingenious work-arounds that users and developers had crafted to meet their EJ goals in spite of the cosmographic constraints imposed by the history and media ecology of GFW.

After observing and interviewing users, we concluded that the best way to integrate global and local views of forests in a non-reductive way was through something we call *storyworld networking*. The term storyworld is derived from a literal translation of 'cosmology,' that is, a story about what the world is like and how we should live in it; networking, we got from theories reviewed in Chapters Two, Three, and Five by Deleuze and Guattari, Latour, Haraway, and others about how to articulate people, events, and stories without reducing them to each other—much like trees articulate with each other to create a coherent, living forest.[12] Using our storyworld networking criteria, we are able to offer in Chapter Five both general and specific recommendations for producers and users of global forest visualization platforms like GFW—and global climate visualization platforms in general. We have also included as an appendix a handlist of forest visualizations that we think build the most compelling storyworld networks.

Why the emphasis on stories? Why do they matter so much to forest conservation and EJ? Stories are likely one of the first human inventions; our species has even been described as the 'storytelling animal.'[13] *Here's why we do it this way … . Here's how we came to be here … . Here's why you should stay out of that forest …* . Stories motivate us when tables, facts, and figures won't. Research has shown that they are an effective driver of EJ action—when people hear each other's stories, what they're going through, they want to help.[14] Cosmograms like the green marble tell stories whether or not we want to listen to them. It is our hope that the storyworld networking criteria we formulate in this project will generate more just, equitable, and compelling visual stories about global forests.

Notes

1 Birgit Müller, personal communication, September 13, 2017.
2 "Restoring Forests, The Lungs of the Planet," *United Environment Program World Conservation Monitoring Centre*, 2021, accessed August 12, 2023, www.

unep-wcmc.org/en/news/restoring-forests--the-lungs-of-the-planet; D. Turner, *The Green Marble: Earth System Science and Global Sustainability* (Columbia University Press, 2018), https://books.google.com/books?id=d0pBDwAAQBAJ.

3 See, for example, Saffron O'Neill and Sophie Nicholson-Cole, "'Fear Won't Do It': Promoting Positive Engagement With Climate Change Through Visual and Iconic Representations," *Science Communication* 30, no. 3 (March 1, 2009), https://doi.org/10.1177/1075547008329201, http://scx.sagepub.com/content/30/3/355.abstract; Stephen R. J. Sheppard, *Visualizing Climate Change: A Guide to Visual Communication of Climate Change and Developing Local Solutions* (London: Routledge, 2012).

4 Rona A. Dennis, et al., "Fire, People and Pixels: Linking Social Science and Remote Sensing to Understand Underlying Causes and Impacts of Fires in Indonesia," *Human Ecology* 33 (2005).

5 M. Walter, et al., *Indigenous Data Sovereignty and Policy* (Taylor & Francis, 2020), 1–20, https://books.google.com/books?id=yBAHEAAAQBAJ.

6 Ben F. Barton and Marthalee S. Barton, "Modes of Power in Technical and Professional Visuals," *Journal of Business and Technical Communication* 7, no. 1 (January 1, 1993), https://doi.org/10.1177/1050651993007001007, http://jbt.sagepub.com/content/7/1/138.abstract; Michel Foucault, "'Panopticism' from 'Discipline & Punish: The Birth of the Prison,'" *Race/Ethnicity: Multidisciplinary Global Contexts* 2, no. 1 (Autumn, 2008); Donna J. Haraway, *Modest_Witness@Second_Millennium.FemaleMan(c)_Meets_OncoMouseTM: Feminism and Technoscience* (New York: Routledge, 1997).

7 Zachary Horton, *The Cosmic Zoom: Scale, Knowledge, and Mediation* (Chicago, IL: University of Chicago Press, 2021); Bruno Latour, "Anti-Zoom" (Contact, catalogue de l'exposition d'Olafur Eliasson, Paris, Fondation Vuitton, 2014); Jussi Parikka, et al., *A Geology of Media* (Minneapolis: University of Minnesota Press, 2015), www.jstor.org/stable/10.5749/j.ctt13x1mnj.

8 Paul Robbins, "Fixed Categories in a Portable Landscape: The Causes and Consequences of Land-Cover Categorization," *Environment and Planning A* 33, no. 1 (2001), https://doi.org/10.1068/a3379, http://journals.sagepub.com/doi/abs/10.1068/a3379.

9 Mike Hulme, "Reducing the Future to Climate: A Story of Climate Determinism and Reductionism," *Osiris* 26, no. 1 (2011).

10 Lisa Westholm and Seema Arora-Jonsson, "Defining Solutions, Finding Problems: Deforestation, Gender, and REDD+ in Burkina Faso," *Conservation and Society* 13, no. 2 (2015).

11 Kharisma Priyo Nugroho, "Rethinking Climate Crisis Solutions in Asian Cities by Harnessing Local Evidence," *PLoS Climate* 2, no. 7 (2023); Nathalie Seddon, et al., "Understanding the Value and Limits of Nature-Based Solutions to Climate Change and Other Global Challenges," *Philosophical Transactions of the Royal Society B* 375, no. 1794 (2020).

12 See note 6 above and Gilles Deleuze and Félix Guattari, *A Thousand Plateaus: Capitalism and Schizophrenia* (London: Bloomsbury Publishing, 1988).

13 Jonathan Gottschall, *The Storytelling Animal: How Stories Make Us Human* (Boston: Houghton Mifflin Harcourt, 2012).

14 Ryan P. Kelly, Sarah R. Cooley, and Terrie Klinger, "Narratives Can Motivate Environmental Action: The Whiskey Creek Ocean Acidification Story," *AMBIO* 43, no. 5 (September 1, 2014), https://doi.org/10.1007/s13280-013-0442-2; Anne Kitchell, Erin Hannan, and Willett Kempton, "Identity Through Stories: Story Structure and Function in Two Environmental Groups," *Human Organization* 59, no. 1 (2000).

Bibliography

Barton, Ben F., and Marthalee S. Barton. "Modes of Power in Technical and Professional Visuals." *Journal of Business and Technical Communication* 7, no. 1 (January 1, 1993): 138–62, https://doi.org/10.1177/1050651993007001007, http://jbt.sagepub.com/content/7/1/138.abstract.

Deleuze, Gilles, and Félix Guattari. *A Thousand Plateaus: Capitalism and Schizophrenia*. London: Bloomsbury Publishing, 1988.

Dennis, Rona A., Judith Mayer, Grahame Applegate, Unna Chokkalingam, Carol J. Pierce Colfer, Iwan Kurniawan, Henry Lachowski, et al. "Fire, People and Pixels: Linking Social Science and Remote Sensing to Understand Underlying Causes and Impacts of Fires in Indonesia." *Human Ecology* 33 (2005): 465–504.

Foucault, Michel. "'Panopticism' from 'Discipline & Punish: The Birth of the Prison.'" *Race/Ethnicity: Multidisciplinary Global Contexts* 2, no. 1 (Autumn, 2008): 1–12.

Gottschall, Jonathan. *The Storytelling Animal: How Stories Make Us Human*. Boston: Houghton Mifflin Harcourt, 2012.

Haraway, Donna J. *Modest_Witness@Second_Millennium.Femaleman(C)_Meets_Oncomousetm: Feminism and Technoscience*. New York: Routledge, 1997.

Horton, Zachary. *The Cosmic Zoom: Scale, Knowledge, and Mediation*. Chicago, IL: University of Chicago Press, 2021.

Hulme, Mike. "Reducing the Future to Climate: A Story of Climate Determinism and Reductionism." *Osiris* 26, no. 1 (2011): 245–66.

Kelly, Ryan P., Sarah R. Cooley, and Terrie Klinger. "Narratives Can Motivate Environmental Action: The Whiskey Creek Ocean Acidification Story." *AMBIO* 43, no. 5 (September 1, 2014): 592–99, https://doi.org/10.1007/s13280-013-0442-2.

Kitchell, Anne, Erin Hannan, and Willett Kempton. "Identity Through Stories: Story Structure and Function in Two Environmental Groups." *Human Organization* 59, no. 1 (2000): 96–105.

Latour, Bruno. "Anti-Zoom." Contact, catalogue de l'exposition d'Olafur Eliasson, Paris, Fondation Vuitton, 2014.

Nugroho, Kharisma Priyo. "Rethinking Climate Crisis Solutions in Asian Cities by Harnessing Local Evidence." *PLoS Climate* 2, no. 7 (2023): e0000255.

O'Neill, Saffron, and Sophie Nicholson-Cole. "'Fear Won't Do It': Promoting Positive Engagement With Climate Change Through Visual and Iconic Representations." *Science Communication* 30, no. 3 (March 1, 2009): 355–79, https://doi.org/10.1177/1075547008329201, http://scx.sagepub.com/content/30/3/355.abstract.

Parikka, Jussi, N. Katherine Hayles, Peter Krapp, Rita Raley, and Samuel Weber. *A Geology of Media*. Minneapolis: University of Minnesota Press, 2015, www.jstor.org/stable/10.5749/j.ctt13x1mnj.

"Restoring Forests, the Lungs of the Planet." *United Environment Program World Conservation Monitoring Centre*, 2021, accessed August 12, 2023, www.unep-wcmc.org/en/news/restoring-forests--the-lungs-of-the-planet.

Robbins, Paul. "Fixed Categories in a Portable Landscape: The Causes and Consequences of Land-Cover Categorization." *Environment and Planning A* 33, no. 1 (2001): 161–79, https://doi.org/10.1068/a3379.

Seddon, Nathalie, Alexandre Chausson, Pam Berry, Cécile A. J. Girardin, Alison Smith, and Beth Turner. "Understanding the Value and Limits of Nature-Based Solutions to

Climate Change and Other Global Challenges." *Philosophical Transactions of the Royal Society B* 375, no. 1794 (2020): 20190120.

Sheppard, Stephen R. J. *Visualizing Climate Change: A Guide to Visual Communication of Climate Change and Developing Local Solutions*. London: Routledge, 2012.

Turner, D. *The Green Marble: Earth System Science and Global Sustainability*. Columbia University Press, 2018, https://books.google.com/books?id=d0pBDwAAQBAJ.

Walter, M., T. Kukutai, S. R. Carroll, and D. Rodriguez-Lonebear. *Indigenous Data Sovereignty and Policy*. Taylor & Francis, 2020, https://books.google.com/books?id=yBAHEAAAQBAJ.

Westholm, Lisa, and Seema Arora-Jonsson. "Defining Solutions, Finding Problems: Deforestation, Gender, and Redd+ in Burkina Faso." *Conservation and Society* 13, no. 2 (2015): 189–99.

2 Forest Maps

The Datafication of Forests From a Media Theory Perspective

In the age of the Internet of Things, it has become normal to digitally connect trees. The city of Melbourne has tagged each of the 70,000 trees growing in its streets with a web address. A map shows which species grow where, the approximate age of individual trees, and their health status. For citizens of the city, it is possible to write an email to each tree. The lives of around 70,000 trees from the Arnold Arboretum at Harvard University in Boston can also be researched via the digital archive BG-BASE on the website http://lifeanddeathofdata.org/. The platform mundraub.org (meaning 'mouth robbery') has marked a map of Germany with trees that bear fruit and are on public land to enable citizens to harvest them at harvest time. Trees in parks and along roads have been recorded in a tree cadaster in Germany since the 19th century so that they can be easily managed. The location, planting age, and tree species are recorded in the cadaster. From 2015 to 2016, New York had the 'TreesCount!' project, a census in which 2,241 volunteers mapped 666,134 trees in the streets and thus measured the high value of trees for the city.[1]

Other platforms do not digitize individual trees with addresses but use the view from above via satellites to analyze forests worldwide. The phrase 'green marble' refers to a NASA map that uses satellite data to analyze the Earth's tree cover and make it visible as a green belt around the Earth (Figure 2.3). Global Forest Watch (GFW), in turn, allows the state of forests to be analyzed globally (Figure 2.2). There are also a lot of national projects which are based on satellite maps to map forests and other types of land. If mapping projects detect forests through satellites, they automatically detect different tree species based on the color, density, pattern, and reflectance of their canopies. They can also determine the height and growth rate of trees from space, and much more. Consequently, satellite detection does not differentiate trees on the basis of local knowledge or plant taxonomy in the tradition of Western botany but on the basis of image and pattern analysis.

In this chapter, we will outline the history and media-technical conditions of digitalization of global forests. How did trees become data? What becomes visible? What is lost in this process? And in which economic tradition do such surveying and inventory projects stand?

DOI: 10.4324/9781003376774-2

From Commodification to Capitalization: A Brief History of Forest Mapping

The following is the self-definition of the global online forest mapping service Global Forest Watch (GFW):

> It's hard to manage what you can't measure. Global Forest Watch makes the best available data about forests available online for free, creating unprecedented transparency about what is happening in forests worldwide. Better information supports smarter decisions about how to manage and protect forests for current and future generations, and greater transparency helps the public hold governments and companies accountable for how their decisions impact forests. GFW data is accessed daily by governments, companies, civil society organizations, journalists, and everyday people who care about their local forests.[2]

Management and surveying, control, forest protection, and knowledge cannot be separated. This relationship has functioned as the reason for creating forest maps since their inception. 'Count it, change it, scale it,' a motto of the World Resources Institute (WRI), GFW's parent organization,[3] is the paradigm and quintessence of any forest management.

European cartographic history has known detailed forest maps since at least the 18th and 19th centuries. This is when the method of 'forestry taxation' was established, which is closely connected to cartographic practices. To understand the different methods used to transfer trees and forests into data (on maps) throughout history, a brief look at the history of cartography is helpful. Today's digital forest maps are still connected to this history, much of which is rooted in Europe. 'Forest mapping was embraced early by emerging European states, first for establishing political boundaries and later for management,' sociologist Nancy Peluso writes, and she emphasizes that '[m]apping of forest resources is therefore an intrinsically political act: whether drawn for their protection or production, they are drawings of a nation's strategic space.'[4] Already the early forest maps from the 18th century established a view from above that detached the view from the ground. This knowledge is intrinsically related to power; as anthropologist James C. Scott writes: '[A]n overall, aggregate, synoptic view of a selective reality is achieved, making possible a high degree of schematic knowledge, control, and manipulation.'[5]

German forest maps can be used as an example to illustrate the relationship between forest mapping and forest management. An early German handbook of nascent forestry science entitled *Instruction for the taxation of forests, or for determining the timber yield of forests* (1795) tells about the problems that exist around the world today as if using a focusing glass 230 years ago.[6] Not only taxation and rights of use such as timber extraction and (often aristocratic) hunting rights but also the distinction among communal, private, manorial, and national ownership, and different forest types are at the heart of the early forest maps. A map from the manual shows how the different forest types are visually distinguished (Figure 2.1). Conifers appear on the map as gray-blue areas, birch trees are pink, and beeches green. Plantations, which don't appear on the exemplary map, are marked by a regular pattern of dots. There are also icons for different tree species and other icons to signify different types of ground, such as sandy soil, clay, or swamp. Boundary signs, including boundary stones and boundary oaks, reveal different properties and types of use. Other icons differentiate forms of housing, including villages, 'head forester's residency' and 'sub-forester's residency,' farm land, gardens, pig farming, and different path types or roads.

In many cases, such maps were part of the land register maps. By the end of the 18th century, the landowners as well as state officials were already using the current methods of land surveying in order to be able to plan their harvests. In the early manuals, the surveying techniques are outlined in detail. Using different colors, signs, and numbers, foresters were to map which trees were planted or grew naturally in which areas. The early forest maps were forest management maps. To become operational, most of them were drawn in large scale (approximately 1:10,000–1:25,000).

Media historian Lisa Cronjäger has shown in detail, following an argument of anthropologist James C. Scott, how new methods of forestry taxation,

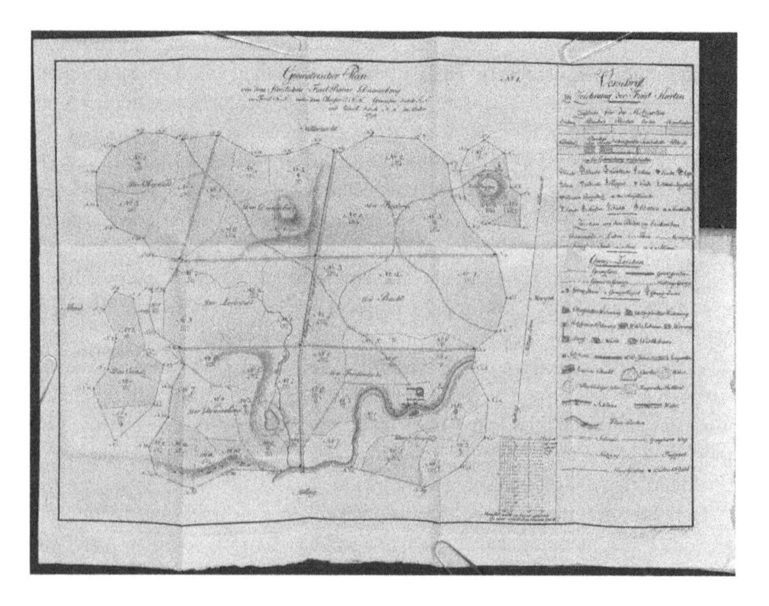

Figure 2.1 Georg Ludwig Hartig: Instruction for the taxation of forests or for determining the timber yield of forests (Anweisung zur Taxation der Forste, oder zur Bestimmung des Holzertrags der Wälder, Giessen, Darmstadt: Heyer, Stahl, 1795).

Source: Figure adapted from Munich, Bavarian State Library, creative commons.

the idea of an optimal rotation age, and forest maps led to the modern idea of forest planning and monocultures—also, how the traditional treatment of forests as a common was repressed and finally criminalized in Europe during the 19th century.[7] The systematic displacement of traditional forms of use in the name of ownership and economic profit, a pattern that has also been operational since colonialism in the Global South, had thus already taken place in Europe since the 18th century. 'The German forest became the archetype for imposing on disorderly nature the neatly arranged constructs of science,' Scott writes, and he develops this idea further: the logic of this modern Western script, to put it bluntly, was based on the reduction of forests and their diverse ecologies to 'timber,' whereby nature became a 'natural resource.'[8] Scott even compares forestry to military hierarchies:

> The forest trees were drawn up into serried, uniform ranks, as it were, to be measured, counted off, felled, and replaced by a new rank and file of lookalike conscripts. As an army, it was also designed hierarchically from above to fulfill a unique purpose and to be at the disposition of a single commander. At the limit, the forest itself would not even have to be seen; it could be 'read' accurately from the tables and maps in the forester's office.[9]

The 'Plantationocene'—the age of the plantation—a term proposed by Donna Haraway and Anna Tsing to oppose the geologic era of the Anthropocene, has its origin in the methods of forest mapping. The methods abstracted forest worlds into extractable natural resources by transforming them into manageable boundary lines and numbers.

Plantationocene: The term was proposed to describe the Anthropocene in more political terms. It describes the spread of the global capitalist logic of standardization, homogeneity, modernization, abstraction, control, extraction, exploitation, and colonial powers, which transform environments into plantations. Typically plantations replace the original biota with monocultures planted in rows and grids, and typically plantations are owned by companies that suppress the traditional land uses of local communities.

At the heart of the optimal rotation age is the ideal of sustainability, which means that cleared forest areas should always be compensated by regrowing forest.[10] 'For this purpose, forest areas are divided into sections in order to cut them down and reforest them one after the other in a cycle principle after a certain number of years has passed.'[11] There is an educational map from the year 1767[12] which shows a forest area divided into 100 numbered rectangles. By this grid, the felling area is ordered into an annual clock; other maps suggested about 80 or even 150–200 years of rotation for oak trees used for shipbuilding. Each year the forester could clear one of the felling areas completely, starting in year one with area 1 and replant new trees. At this point, one could say, the logic of monoculture imposed itself because timber is much easier to plan and calculate, if only pines or spruces are planted at this time, because they grow quickly and in tandem. Such principles were introduced in the context of forestry, showing owners and landlords their forests as timber resources and offering a planning scheme at the same time. The flip side of this practice was that other forest uses were devalued in favor of timber as 'forest minor uses.' Thus, it can be generalized that forest maps and the bureaucratization of forest taxation methods exacerbated these practices, which had been in use long before detailed maps were drawn in areas where a lot of timber was needed close to cities or mines.

The critique and rhetoric of 'unsustainable forest management' was already inscribed on the early forest maps as an argument for modernization. Indeed, it had been a concern for centuries already. As environmental historian Joachim Radkau notes, it is cited as the reason for the sometimes only alleged lack of concern of the various forest users for ever new maps. The preface to a 1797 manual alludes to forest development in Germany with what were then commonplaces. The forest that Roman historian Tacitus impressively described in his 1st-century book *Germania* had long since been cleared, as the manual's author, a forest district manager in the Fichtel Mountains, writes. Instead, 'insatiable greed' had 'pushed wild free nature up the mountains' to satisfy the need for wood. Greed is considered by the author to be the 'ruling basic instinct of human nature,' especially of those who own forest or arable land.[13] This could be limited solely by regulations from above, the forest manager said.

Michael Williams, in his 2006 book *Deforesting the Earth*, told the global forest history as a single deforestation story. Radkau critically notes that a forest history informed solely by forest records 'automatically becomes a deforestation history,'[14] but that it is actually more complex because of the many conflicting goals of forest users and the numerous types of use. The trajectory would not always go in the direction of deforestation regionally. Regionally, this may be true, but a look at global numbers contradicts this argument, as here forest losses exceed gains and the 'gains' often consist of new plantations. And yet, the issue of timber scarcity has been an important driver for more forest management throughout history in the paradigm of loss control; indeed, timber shortage as a rhetoric figure runs through history to enforce goals and eventually led to forest economics.[15]

Forest economics assesses and calculates timber masses in terms of their annual increment or loss; it is based on forest yield theory, which views forests as capital. Methods gathered in early forestry textbooks included the introduction of measuring instruments such as dendrometers, which could be used to measure the height of trees or determine their diameter and more. But the different tree species, their quality, and yield also played a significant role, leading to the large-scale planting of coniferous forests in Germany, where mixed or deciduous forests had previously been native. Added to this were the techniques of land surveying, which became much more accurate in the 19th century using the method of triangulation. One further reason why alongside taxation also came the preference for monocultures rather than mixed forests is that mixed forests, like primeval forests, were much more difficult to calculate than simultaneously planted, straight-growing plantations.

Another general effect of maps consists in their 'naming power.' In the context of the mapping of Switzerland during the 19th century, historians David Gugerli and Daniel Speich speak of how the 'paper surfaces' of maps already exert a 'naming power on the designated spaces whose consequences are often reality-shaping.'[16] This designating power also applies to forest maps. On all maps, it must be decided what places, rivers, and mountains are named. Multiple names by natives in different languages or dialects, as was the case in Switzerland at the time, are overwritten in favor of a single, 'official' name. On a much larger scale, indigenous naming systems in the colonies were mostly erased in this process. This is also true for the naming systems of tree species that follow the language of the conquerors on the maps.[17] What counts as forest and what type of forest is classified must be standardized for the maps. This goes hand in hand with the processes described by anthropologist Jack Goody in his 1977 book *The Domestication of the Savage Mind.* Lists, tables, and writing, in general, change ontology. In the case of maps, all ontologies must be adapted to the map. There is no room for pluriverses on maps belonging to individuals in power—no space for worlds other than 'the civilizational model,' as Arturo Escobar has argued.[18]

As indicated earlier, parallel to the development of forest maps, the right of ownership and use changed radically. The changes had much to do with the emergence of the modern concept of property, which has since conflicted with old customary rights such as plenter forest ('Plenterwald'), coppice ('Niederwald'), and hud forest ('Hudewald') for house building, firewood, and livestock. The development of forest maps in Europe has been linked to the conflicts that arose due to ownership and use rights. The question of ownership according to this model was also transferred to the colonies. Forest regulations and forest protection against the multiple cooperative uses of forests and the devaluation of these uses as secondary uses can be enforced with maps. The history of forests is a history of forests and power, especially because in the 'wooden culture'[19] (Werner Sombart, 1902) wood was the only burning material available before the age of lignite, coal, and oil. Taking Germany as an example again, we can see how during the course of the 19th century the centuries-old right of allowing local residents to take wood lying around in the forest for their stoves was criminalized as 'wood theft' ('Holzfrevel'). Henceforth, those who nevertheless continued to exercise this right had to be punished. To this day, only branches a few centimeters thick may be taken from the ground in many European forests. These are the conflicts in which forest maps unfold their power of definition by assessing what grows where and designating who owns particular parcels of land. (It is no coincidence that the radical changes in forest management and forest rights were the starting point for the social critique of Karl Marx, when in 1842, as a young editor for the *Rheinische Zeitung*, he wrote a critical article about the new wood theft law. It was the cause of his dismissal from the newspaper, but also the beginning of his thinking about the social effects of ownership.)

Forest maps gain their importance in a context of scarcity, entitlement, and ownership. For in part, the maps were a reaction to the years of timber scarcity; they were a means of planning timber management in a more sustainable way, that is, a means of forest management that allowed even more timber to be extracted from a forest in the long term. By 1800, the goal was to simultaneously protect forests and make them more productive through better methods. Otherwise, as the author of the 1797 forestry manual states, more families would have to emigrate to America to cut timber. The maps made it possible to see at a glance the stocks and thus the potential of land. In this way, however, they were always also a means for the development and future exploitation of resources. Maps have stood in this paradox from the very beginning. They are a medium of protection and exploitation at the same time.

Peluso defines a forest map as a medium 'for contesting the homogenization of space on political, zoning, or property maps, for altering the categories of land and forest management, and for expressing social relationships in space rather than depicting abstract space in itself.'[20] This definition applies equally to analog and digital maps. Even though mapping methods have changed, current platforms such as GFW follow the tradition of forest management maps. The paradigm of the forest map applies to both digital and analog maps: both place a taxonomy over the Earth, showing which resources are where and which different forest types exist where. Both identify legal relationships, ownership, nation-states, and protected areas. Global forest map platforms such as GFW have to work both with and against this power.

Mapping, Globalization, Action: From Blue Marbles to Google Engines

The history of mapping has undergone a radical technical transformation with the advent of photography of the Earth from space, and at the same time, the basic features developed in analog maps continue to operate in digital maps. Everyone reading this book will be familiar with the image of the Earth from space, calling in particular for the protection of the Earth as a habitat. For 50 years, the blue marble has been the strongest image of any environmental movement. It's no exaggeration to say that the photo shot on the flight to the Moon became the icon that visually united all 'earthlings' because its message was so simple, beautiful, and evident: Earth is a unique living planet with a dynamic atmosphere in the solar system, it is a global ecology of relations, and it is the basis and prerequisite for (human) life.[21] Similarly, readers might be familiar with 'red marble' or 'burning world' images that appeared alongside the UN Intergovernmental Panel on Climate Change (IPCC) reports from the 1990s and following—visualizations of the Earth's surface temperature under various global warming projections in different shades of red. Not many readers may be familiar, however, with the 'green marble' images, which, thanks to recent advances in remote sensing and digital mapping, have enabled a vision of the global Earth as a function of its vegetation, so also the coverage of the Earth with forests (Figure 2.2).

Today, we know that the power of the blue marble image turned into a hypocritical image of nature. Although the image was used time and again as an impulse to unite actors, profound subsequent actions 'to save the planet,' as the slogan goes, have not yet been taken. Green marble images have been developed in response to this criticism as a vehicle to give the blue marble subjectivity and agency. With the launch of the US *Landsat* program in 1972, just when the 'ecological turn' was reaching its peak, surface vegetation and forest cover became an important focus of satellite observation.[22]

The recognition of different types of forest cover with the means of satellite investigation has been an interest of different stakeholders—ecologists, climate scientists, and forest managers alike. They are interested in how much land surface is covered with forests and how surface vegetation changes throughout the year and in longer time periods. In 2011, thanks to a system sensing visible and infrared wavelengths installed on the NOAA/NASA weather satellite Suomi NPP, it became possible to monitor global forest dynamics. These satellite instruments help observe weather, climate, oceans,

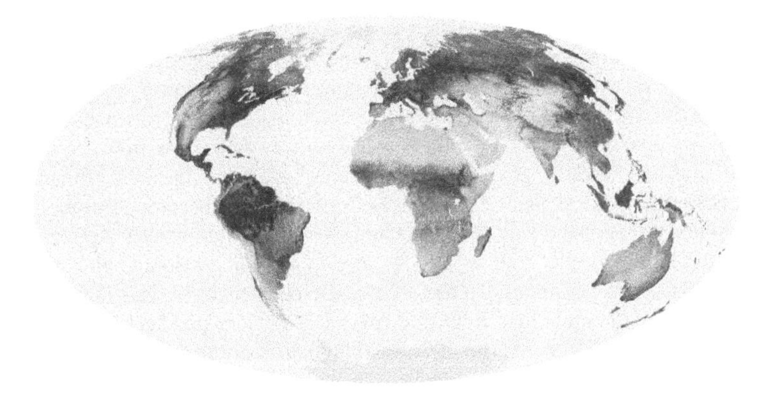

Figure 2.2 'Green marble'—global vegetation as seen by Suomi NPP, NASA/NOAA 2013.

Source: Figure adapted from NASA/NOAA.

nightlight, wildfires, movement of ice, and changes in vegetation and landforms. In 2013, the measurement series of Suomi NPP resulted in the 'green marble' made from thousands of composite images. Like a thick carpet, a green layer covers large parts of the continents, while the other parts of the Earth appear in white, the color of the 'terra incognita,' as if there were no vegetation there at all. This contrast makes the forests of the Earth even more visible. The climatic zones of the tropics and temperate latitudes are clearly visible, while the dry deserts stand out from them with their light beige color. By means of this kind of representation, the Earth appears as a living and potentially growing organism, like a colony of algae or moss.

Nevertheless, green marble images are still undeniably the product of the military-industrial complex, of which satellite imagery emerged, and the politics flowing out of that complex—because satellites were always launched to spy on other powers. This logic also continues to work. Forests are monitored as a matter of biosecurity and economic profit: to that extent, the criticisms of blue marble images also apply to green marble ones—that is, they depict the Earth as a resource to be capitalized on by transnational neoliberal corporations and programs.[23] The response to this criticism has been to complement global perspectives of the Earth's environment and climate with local ones. Multiple climate-monitoring platforms—such as Global Climate Monitor, the EUMETSAT platform, and NASA and NOAA platforms—have made extensive use of whole Earth images and maps expressly in order to 'downscale' marble-type images of the Earth into local perspectives, a development that will be treated in more depth in the following chapter.

The green marble has been computerized as well: forest data captured by satellites have been integrated with GPS/GIS mapping platforms—primarily Google Earth—in order to provide interactive functionality such as zooming, searching, and tagging. Also, climate action and climate services meet in these platforms because global data maps, satellite images, and visualizations are essential means for enabling more responsible forms of interaction with the planet. Indeed, interactive green marble platforms had been quickly adopted by REDD+ programs ('Reducing Emissions from Deforestation and Forest Degradation'). Since forests are one of the most brilliant transformers of CO_2 into O_2, platforms like GFW have become ideal tools of the transnational organizations that now serve as Anthropocene 'stewards of the Earth system,'[24] measuring the forest as a living resource globally, calculating its oxygen production as a single number. Users can calculate carbon dioxide emissions from tree cover loss by tons per hectare on regional and global levels. The green marble maps share the ambivalence of care and control, just as it has already been emphasized for forest maps in the previous section.

The global view of global forest mapping platforms is articulated by both the media ecology of the Google Earth Engine, which is the basis of platforms like GFW, and the ecology of the forest (Figure 2.3). Google defines its engine as a combination of 'a multi-petabyte catalog of satellite imagery and geospatial

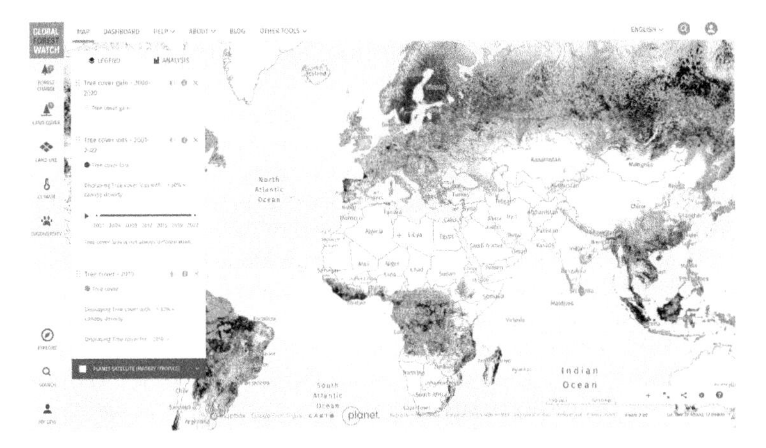

Figure 2.3 The landing page of the platform Global Forest Watch run on the basis of Google Earth Engine. Forest areas are indicated in dark grey and forest loss is marked in light grey.

Source: Screenshot from http://www.globalforestwatch.org/, August 15th 2023.

datasets with planetary-scale analysis capabilities.'[25] The company makes its data ecological engine available for scientists, researchers, and developers 'to detect changes, map trends, and quantify differences on the Earth's surface.'[26] On the website of the WRI, Andrew Steer, the president and CEO, is prominently quoted as saying, 'Google Earth Engine has made it possible for the first time in history to rapidly and accurately process vast amounts of satellite imagery, identifying where and when tree cover change has occurred at high resolution. Global Forest Watch would not exist without it. For those who care about the future of the planet, Google Earth Engine is a great blessing!'[27] The global view of satellite imagery and datasets via the Google Engine tool is really the only way to experience global forest change. The dominance of Google leads to the question: how are power and knowledge connected—if vision and action on global forests must be linked within the logic of this engine? The Earth from above and the view of forests are public and universal. For forest views, there is no right, as with Google Street View, to deny the view from above and hide it with a blurred gray cloud. Only certain military areas are entitled to exercise this right. If clouds interfere with the view from the sky, they are later replaced by shots taken when the sky is clear. The promise of Google Earth and satellite imagery is the ability to see everything, right down to the last corner of the Earth.

How are seeing, knowing, and acting linked in such global mapping tools? The Green marble is in many ways a combination of the blue and red marbles: it presents the Earth as a coherent, living organism but also communicates its fragile, threatened situation. Furthermore, through its incorporation into interactive visualization platforms for monitoring deforestation like GFW, the green marble promises something the blue and red marbles could not—intervention.

'Count It, Change It, Scale It': the slogan that announces the use of GFW relates measurement to (political) action. Digital mapping in the case of forests is based on the assumption of 'actionable data.' To make data actionable, one needs assemblages of mapping, GPS, satellite observation, and online platforms such as GFW. In the previous section, we have already seen how forest maps became a central method of transforming global forests into resources and plantations that can be controlled. This idea can be transferred to satellite images.

To understand how images can initiate action and to what extent images are an integral component for action, it can be helpful to draw on the idea of operational or operative images—both spellings are in use. Filmmaker and author Harun Farocki introduced this term to reflect on photographs or film shots 'taken from a position that a person cannot normally occupy.' The special property of operative images is that they 'do not represent an object, but

Operative/operational images: Operational images are instruments that make it possible to do things in the world, that is, to plan and control action. In this way, images become instruments or techniques that enable action. Operational images make something visible and recognizable, while at the same time helping to render these things controllable, operable, and manageable.

rather are part of an operation.'[28] The term can be applied illuminatingly to the application of interactive forest maps such as in GFW, which are based on interactive online maps of satellite imagery and which become active tools when people start to use them in a context of action. Operative images such as maps based on satellite images are posthuman representations to a large extent because the largest parts of the maps are automatically classified by algorithms.[29]

When images become operational, top-down control and bottom-up empowerment can be equally linked to them. GFW's slogan 'forest monitoring designed for action' aims precisely in this direction. Planetary images such as the blue marble and the green marble, and different applications of the Google Earth Engine and global satellite services can be seen as forensic tools and operational images which since their inception contained a call for action. This is because public satellite imagery serves as a forensic exhibit to which everyone can point to make their case against transgressors. *Forensis*, as Eyal Weizman from the collaborative research agency Forensic Architecture calls it, is 'Latin for "pertaining to the forum,"' originally a broad space of politics, law, and economy and 'its potential as a political practice.'[30]

> **Forensis/forensic methods:** A critical practice that investigates the hidden actions of states and powerful corporations on the basis of mixed methods, including mapping practices, satellite observations, and investigations in situ, to make them public or bring them to court.

In modernity, 'the forum gradually came to refer exclusively to the court of law, and forensics to the use of medicine and science within it. . . . Things too far away, too abstract, or too large—such as cities, resources, rivers, territories or states—had to be brought vividly to life by the power of an aural demonstration.'[31] Forensic science in combination with satellite imagery and other forensic research methods can turn into a critical action that reveals hidden actions. Forensic Architecture, for example, uses mapping methods and satellite data among many other sources to investigate human rights violations by states, police forces, militaries, and corporations. For instance, they have analyzed ecocides in Indonesia, drone strikes in Pakistan, airstrikes on a hospital in Syria, intentional fires in Papua, and gold mining in the Amazon Rainforest.[32]

This is the other side that comes into effect when global satellite images become available to the general public and thus for civilian purposes. When this occurs, the images can also be turned against those in power. Forensic methods render visible those hidden or invisible processes that have so far gone unnoticed. Cindy Lin, in her chapter on forest mapping by means of satellites in Indonesia, compares the willful invisibility with 'Elias Canetti's dictum that secrets are at the heart of power,' referring to Michael Taussig.[33] Measuring and mapping forests from the sky is one way to make hidden actions visible by investigating tiny traces of changes in the texture of a landscape that is openly observable from the sky. Satellite imagery was first installed for military use and enemy surveillance before it was made public for civilian use. This military provenance hints at the tactical and strategic logic of satellite-powered maps which is still in their DNA. Since becoming public, these eyes can also be used strategically by less powerful actors for civilian purposes. Even though everyone must be aware that the old forces remain inseparable from the tools, this is a form of power and potential that lies in forest map platforms such as GFW—that is, the ability to counter-surveil and, in doing so, to act against those in power.

Google Warming and Google Gaia

We have already pointed out that the global view provided by forest mapping platforms is articulated by both the media ecology of the Google Earth Engine and the ecology of the forest. In the following section, we want to take the idea of media ecology in the literal sense even further.

Following John Tresch's work, one of us (Schneider) has already labeled the red and blue marbles as moralistic images of the Earth or 'cosmograms.' Cosmograms tell stories about how everything within them should be seen as working together. They are images of the world that 'establish the relation between different domains or ontological levels.'[34] In an interview, anthropologist John Tresch explains further:

> All cultures have cosmograms, which are attempts to say: 'This is how the world works, this is how everything fits together'—humans, all the divisions of nature, all the divisions with human society, and then the divinities around it or above it, the metaphysics underlying it. In order to convey cultures and beliefs, to teach them, to re-inscribe them and make them true and activate them, they need some kind of form to embody them. And I call anything that takes that form a cosmogram. It can be a building, a painting, a poem, or a book like the Bible—or a song. It can apply to many, many different kinds of human products.[35]

A cosmogram functions as a synopsis of a cosmology, which Tresch defines as follows:

> A cosmology is more than a system of classification, an origin myth, or a theory of the relationships among what there is in the universe; it also involves affective and aesthetic dimensions and the sense of coherence of a group's characteristic words, practices, and objects.[36]

In this sense, any images showing the cosmic order, such as religious altarpieces, mandalas, and the medieval map we reprinted in Figure 1.1, all count as cosmograms. Marble images of the Earth also count as cosmograms, as do images of the Earth as a coherent, living organism that participates in a larger cosmology often referred to as Gaia. Within the Gaia hypothesis developed by James Lovelock and Lynn Margulis in the 1970s, the Earth is seen as a coherent living whole consisting of uncountable feedback loops. Bacteria, algae, and plants play the major role in the composition and stability of the atmosphere that makes life possible. Even if forests are only a part of this ecology, they are an important contributor when it comes to the production of oxygen and the decomposition of carbon dioxide. Global forests are an essential component of Earth's self-regulating lungs. So, blue and green marbles evoke a Gaian cosmology; red marbles do, too, if we think of them as images of a fevered or ill Gaia.

What happens to Gaia when she is brought inside a computer? Media scholar Leon Gurevitch has thought through this problem for blue marble images in his study of Google Earth. For him, the key transformation here is between the blue marble as a static, analog representation of the Earth and Google Earth as a dynamic mathematical model. This key transformation renders the Earth as a design object, able to be altered with a click of the mouse as is a computer-assisted design (CAD) rendering of a building. He writes, '[n]either entirely virtual nor entirely indexical, Google Earth operates as a machinic hybrid in which the panoptical power of satellite imaging is combined with the simulative capacities of the product design—engineered object.'[37] Furthermore, Gurevitch argues that this design-centered view of Google Earth leads to global engineering solutions to climate change, such as stratospheric aerosol injection or space mirrors. He terms this phenomenon 'Google Warming' and links it to neoliberal, transnational models of economic globalization: 'Representing both the environmental feedback of satellite surveillance and the computer-automated construction of a virtual environment the machinic panopticism of Google Earth reflects a new representational politics in which the Earth's ecosystems are rationalized as always already industrialized (or industrializable).'[38] Both aspects are a central part of the logic shared by global forest-monitoring platforms. For this reason, we have summarized the cosmology of the Earth's forests in which the cosmograms generated by these platforms participate as *Google Gaia.*

Google Gaia: A combination of instrumental feedback loops that constructs the world's forests as a single, living organism, one whose health can be monitored and intervened remotely from a computer.

What does it mean if we think of interactive forest maps powered by the Google Engine as a Google Gaia? It mixes the Gaian biosphere with the human technosphere, exploitative in so many cases, in a feedback loop. But if monitoring platforms such as GFW only represent Gaia processes, it would not be a Google Gaia. To deserve this name, the visualization tool itself must be part of Gaia's feedback loops. And, as we argue, this is indeed the case if we apply this thought to global forest-monitoring platforms: first, the platform is connected to the biosphere on a material level by using up energy and materials to drive all agents of the monitoring infrastructure such as satellites, internet infrastructures and data centers, and server farms. Second, the platform as a monitoring tool is meant to function as a reason to act and by this intervene directly in the techno-biosphere with political measures—in other words, to induce political action.

It seems clear now that forest-monitoring platforms participate in what Leon Gurevitch calls 'Google Warming.' As discussed earlier, this idea suggests that we can control the world with technical solutions once it has been brought inside a computer, as with Google Earth: a keystroke here or there makes alterations to the image; the same kind of leverage applies to ideas for geo-engineering on a global scale to combat climate change—deploying space mirrors or spraying tons of calcium carbonate into the stratosphere. In fact, a transnational nonprofit called SilverLining just announced $3 million in grants to test various stratospheric aerosol injection (SAI) scenarios in global computer models to see how much solar reflection and cooling can be achieved before various negative effects accumulate in extreme weather, ocean pH change, etc. One engineer associated with the tests expressed their primary research question as follows: 'Is there a way, in our model world at least, to see if we can achieve one without triggering too much of the other?'[39] The conflation of Google Earth with the real Earth is stark, and so is the nearly nuclear scale of the interventions and consequences being discussed nonchalantly by white Western researchers who will be buffered by their privilege from the fallout of their scenarios at a distance. One expert quoted by the *New York Times* story on the SilverLining grant compared SAI to 'chemotherapy for the planet,' and a major funder of SilverLining issued the following dramatic statement: 'If we don't explore climate interventions like sunlight reflection now, we are surrendering countless lives, species, and ecosystems to heat.'[40] However, the article fails to mention the input of the 'lives, species, and ecosystems' to mitigating climate change or to consider the impact of SAI itself on them.

Here we find ourselves again on the negative side of cartography and the powers it serves through the tremendous power inscribed in its own logic. If the top-down logic of the platform is totalized, global, technocratic action will be imposed on everyone's environments. Gaia Googlists think they can fix the planet because in their models forest-ecological problems always work out if the parameters are set correctly. Plantations and geo-engineering are both logical solutions to problems which were made visible on the early forest maps and which are visualized from space today. So the logic of the Google Gaia Engine is yet another justification for a blue marble, instrumental stewardship of the Earth.

We aren't the only ones to recognize these problems with globalization of climate and forest visualization—so have EJ advocates, particularly in the Global South. The primary response to globalization been 'glocalization,'[41] or the development of visualization tools that allow zooming in from global to local views of environment for the purposes of analysis—like GFW. We consider the geopolitical implications of glocalization platforms in the next chapter.

Notes

1 Jonathan Gray, "The Datafication of Forests? From the Wood Wide Web to the Internet of Trees," in *Critical Zones: The Science and Politics of Landing on Earth*, ed. Bruno Latour and Peter Weibel (Karlsruhe: ZKM—Center for Art and Media Karlsruhe, 2020), 364–271.
2 "About," *Global Forest Watch*, 2023, accessed August 23, 2023, www.globalforestwatch.org/about/.
3 "About Us," *World Resources Institute*, 2023, accessed August 23, 2023, www.wri.org/about.
4 Nancy Peluso, "Whose Woods Are These? Counter-Mapping Forest Territories in Kalimantan, Indonesia," *Antipode* 274 (1995): 383–406, both quotes 383.
5 James C. Scott, *Seeing Like a State: How Certain Schemes to Improve the Human Condition Have Failed* (New Haven and London: Yale University Press, 1998), 11.

6 Georg Ludwig Hartig, *Anweisung zur Taxation der Forste, oder zur Bestimmung des Holzertrags der Wälder* (Giessen, Darmstadt: Heyer, Stahl, 1795).
7 Lisa Cronjäger, *Umtriebszeiten. Forsteinrichtungskarten und Waldnutzungspraktiken zwischen Nachhaltigkeit und Holzfrevel (1760–1860)* (PhD diss., University of Basel, Switzerland, 2022). English chapter on the subject: Lisa Cronjäger, "On the Exactly Balanced Use of Nature and Its Representation," *History of Humanities, Forum: The Promises of Exactitude* 7, no. 2 (2022): 161–175.
8 Scott, *Seeing Like a State*, 15.
9 Ibid.
10 Paul Warde, *The Invention of Sustainability: Nature and Destiny, c.1500–1870* (New York: Cambridge University Press, 2018); K. Jan Oosthoek and Richard Hölzl, eds., *Managing Northern Europe's Forests: Histories from the Age of Improvement to the Age of Ecology* (New York and Oxford: Berghahn, 2018).
11 Cronjäger, *Umtriebszeiten*, 3 (transl. BS).
12 "Das Revier Freudenberg in ein 100 Jähriges Gehau vertheilet," engraving from the book by Johann Ehrenfried Vierenklee, *Mathematische Anfangsgründe der Arithmetik und Geometrie: in so fern solche denjenigen, die sich dem höchstnöthigen Forstwesen auf eine vernünftige und gründliche Weise widmen wollen, zu wissen nöthig sind* (Leipzig: Weidmann, 1767).
13 Heinrich Christoph Moser, *Die practisch-geometrische Aufnahme der Waldungen mit der Bousole und Meßkette* (Leipzig: Gräff, 1797), v.
14 Joachim Radkau, *Holz: Wie ein Naturstoff Geschichte schreibt* (München: Oekom Verlag, 2018), 27–28.
15 Radkau, *Holz*, 154.
16 David Gugerli and Daniel Speich, *Topografien der Nation. Politik, kartografische Ordnung und Landschaft im 19. Jahrhundert* (Zurich: Chronos Verlag, 2002), 75 (transl. BS).
17 Londa Schiebinger and Claudia Swan, eds., *Colonial Botany: Science, Commerce, and Politics in the Early Modern World* (Philadelphia: University of Pennsylvania Press, 2005).
18 Arturo Escobar, *Designs for the Pluriverse: Radical Interdependence, Autonomy, and the Making of Worlds* (Durham: Duke University Press, 2018).
19 Werner Sombart, *Der moderne Kapitalismus*, Vol. 2, Die Theorie der kapitalistischen Entwicklung (Leipzig: Duncker & Humblot, 1902), 1138.
20 Henri Lefebvre, *The Production of Space* (Oxford: Blackwell Publishing, 1991); Peter Vandergeest and Nancy Lee Peluso, "Territorialization and State Power in Thailand," *Theory and Society* 35 (1995): 385–426.
21 Cf. Sheila Jasanoff, "Image and Imagination: The Formation of Global Environmental Consciousness," in *Changing the Atmosphere: Expert Knowledge and Environmental Governance*, ed. Paul Edwards and Clark A. Miller (Cambridge, MA: MIT Press, 2001), 309–337; Diedrich Diederichsen, Anselm Franke, and Haus der Kulturen der Welt, eds., *The Whole Earth: California and the Disappearance of the Outside* (Berlin: Sternberg Press, 2013); Sebastian Vincent Grevsmühl, *La Terre Vue D'en Haut: L'invention De L'environnement Global* (Paris: Seuil, 2014); Finis Dunaway, *Seeing Green: The Use and Abuse of American Environmental Images* (Chicago: The University of Chicago Press, 2018); Vera Tollmann, *Sicht von oben: "Powers of Ten" und Bildpolitiken der Vertikalität* (Berlin: Spector Books, 2021).
22 Lisa Parks has published extensively on satellite imagery and culture. E.g. *Cultures in Orbit: Satellites and the Televisual* (Durham, NC: Duke University Press, 2005). The term 'ecological turn' was theorized by Patrick Kupper, Joachim Radkau, and Frank Uekötter.
23 Cf. David Humphreys, "Life Protective or Carcinogenic Challenge? Global Forests Governance under Advanced Capitalism," *Global Environmental Politics* 3, no. 2 (2003): 40–55.

24 Paul J. Crutzen and Will Steffen, "How Long Have We Been in the Anthropocene Era?" *Climatic Change* 61, no. 3 (December 1, 2003): 251–57, 256.
25 Cf. Google Engine website.
26 Ibid.
27 Ibid.
28 Harun Farocki, "Phantom Images," *Public. Art, Culture, Ideas* 29 (2004): 12–22, 17.
29 Many authors have since contributed to make this concept fruitful for technical images, such as Volker Pantenburg, Nina Samuel, Kathrin Friedrich, Charlotte Klonk, Jussi Parikka, Jan Distelmeyer, or the artist Trevor Paglen. A good overview gives Aud Sissel Hoel, "Operative Images: Inroads to a New Paradigm of Media Theory," in *Image—Action—Space: Situating the Screen in Visual Practice*, ed. Luisa Feiersinger, Kathrin Friedrich and Moritz Queisner (Berlin: De Gruyter, 2018), 11–27.
30 Eyal Weizman, "Introduction," in *Forensis: The Architecture of Public Truth*, ed. Forensic Architecture (Berlin: Sternberg Press, 2014): 9–32, 9–10.
31 Ibid.
32 https://forensic-architecture.org/
33 Cindy Lin, "How to Make a Forest," *E-Flux Architecture*, 2020, www.e-flux.com/architecture/at-the-border/325757/how-to-make-a-forest/.
34 John Tresch, "Cosmogram," in *Cosmogram*, ed. Melik O'Hanian and Jean-Christophe Royoux (New York: Sternberg, 2005), 67–76 and John Tresch, "Technological World-Pictures," *Isis* 98, no. 1 (2007): 84–99; Birgit Schneider, "Burning Worlds of Cartography: A Critical Approach to Climate Cosmograms of the Anthropocene," *Geo: Geography and Environment* 3, no. 2 (2016): e00027.
35 "Interview with John Tresch," *Sonic Acts*, no. 20 (2015), accessed August 23, 2023, https://sonicacts.com/mobile/portal/interview-with-john-tresch-on-cosmograms.
36 Tresch, "Technological," 84f.
37 Leon Gurevitch, "Google Warming: Google Earth as Eco-Machinima," *Convergence: The International Journal of Research into New Media Technologies* 20, no. 1 (2014): 85–107, 88.
38 Ibid., 88.
39 Christopher Flavelle, "As Climate Disasters Pile Up, a Radical Proposal Gains Traction," *New York Times*, October 28, 2020.
40 Ibid.
41 Edward W. Soja, *Postmetropolis: Critical Studies of Cities and Regions* (Malden, MA: Blackwell, 2000), 199.

Bibliography

"About." *Global Forest Watch*, 2023, accessed August 23, 2023, www.globalforestwatch.org/about/.

"About Us." *World Resources Institute*, 2023, accessed August 23, 2023, www.wri.org/about/.

Cronjäger, Lisa. *Umtriebszeiten. Forsteinrichtungskarten und Waldnutzungspraktiken zwischen Nachhaltigkeit und Holzfrevel (1760–1860)*, PhD diss., University of Basel, Switzerland, 2022.

Crutzen, Paul J., and Will Steffen. "How Long Have We Been in the Anthropocene Era?" *Climatic Change* 61, no. 3 (December 1, 2003): 251–57.

Diederichsen, Diedrich, Anselm Franke, and Haus der Kulturen der Welt. *The Whole Earth: California and the Disappearance of the Outside*. Berlin: Sternberg Press, 2013.

Dunaway, Finis. *Seeing Green: The Use and Abuse of American Environmental Images*. Chicago: The University of Chicago Press, 2018.
Escobar, Arturo. *Designs for the Pluriverse: Radical Interdependence, Autonomy, and the Making of Worlds*. Durham, NC: Duke University Press, 2018.
Farocki, Harun. "Phantom Images." *Public. Art, Culture, Ideas* 29 (2004): 12–22.
Flavelle, Christopher. "As Climate Disasters Pile up, a Radical Proposal Gains Traction." *New York Times*, October 28, 2020.
Gray, Jonathan. "The Datafication of Forests? From the Wood Wide Web to the Internet of Trees." In *Critical Zones: The Science and Politics of Landing on Earth*, edited by Bruno Latour and Peter Weibel, 364–71. Karlsruhe: ZKM—Center for Art and Media Karlsruhe, 2020.
Grevsmühl, Sebastian Vincent. *La Terre Vue D'en Haut: L'invention De L'environnement Global*. Paris: Seuil, 2014.
Gugerli, David, and Daniel Speich. *Topografien der Nation. Politik, kartografische Ordnung und Landschaft im 19. Jahrhundert*. Zurich: Chronos Verlag, 2002.
Gurevitch, Leon. "Google Warming: Google Earth as Eco-Machinima." *Convergence: The International Journal of Research into New Media Technologies* 20, no. 1 (2014): 85–107.
Hartig, Georg Ludwig. *Anweisung zur Taxation der Forste, oder zur Bestimmung des Holzertrags der Wälder*. Giessen, Darmstadt: Heyer, Stahl, 1795.
Hoel, Aud Sissel. "Operative Images. Inroads to a New Paradigm of Media Theory." In *Image—Action—Space: Situating the Screen in Visual Practice*, edited by Luisa Feiersinger, Kathrin Friedrich and Moritz Queisner, 11–27. Berlin: De Gruyter 2018.
Humphreys, David. "Life Protective or Carcinogenic Challenge? Global Forests Governance Under Advanced Capitalism." *Global Environmental Politics* 3, no. 2 (2003): 40–55.
Jasanoff, Sheila. "Image and Imagination: The Formation of Global Environmental Consciousness." In *Changing the Atmosphere: Expert Knowledge and Environmental Governance*, edited by Paul Edwards and Clark A. Miller, 309–37. Cambridge, MA: MIT Press, 2001.
Lefebvre, Henri. *The Production of Space* [in English; French (translation)]. Translated by D. Nicholson-Smith. Oxford: Blackwell Publishing, 1991.
Lin, Cindy. "How to Make a Forest." *E-Flux Architecture*, 2020, www.e-flux.com/architecture/at-the-border/325757/how-to-make-a-forest/
Moser, Heinrich Christoph. *Die practisch-geometrische Aufnahme der Waldungen mit der Bousole und Meßkette*. Leipzig: Gräff, 1797.
Parks, Lisa. *Cultures in Orbit: Satellites and the Televisual*. Durham, NC: Duke University Press, 2005.
Peluso, Nancy Lee. "Whose Woods Are These? Counter-Mapping Forest Territories in Kalimantan, Indonesia." *Antipode* 27, no. 4 (1995): 383–406.
Radkau, Joachim. *Holz: Wie ein Naturstoff Geschichte schreibt*. München: Oekom Verlag, 2018.
Schiebinger, Londa, and Claudia Swan, eds. *Colonial Botany: Science, Commerce, and Politics in the Early Modern World*. Philadelphia: University of Pennsylvania Press, 2005.
Schneider, Birgit. "Burning Worlds of Cartography: A Critical Approach to Climate Cosmograms of the Anthropocene." *Geo: Geography and Environment* 3, no. 2 (2016): e00027.

Scott, James C. *Seeing Like a State: How Certain Schemes to Improve the Human Condition Have Failed.* New Haven and London: Yale University Press, 1998.

Soja, Edward W. *Postmetropolis: Critical Studies of Cities and Regions.* Malden, MA: Blackwell, 2000.

Sombart, Werner. *Der moderne Kapitalismus.* Bd. 2, Die Theorie der kapitalistischen Entwicklung. Leipzig: Duncker & Humblot, 1902.

Tollmann, Vera. *Sicht von oben: „Powers of Ten" und Bildpolitiken der Vertikalität.* Berlin: Spector Books, 2021.

Tresch, John. "Cosmogram." In *Cosmogram*, edited by Melik O'Hanian and Jean-Christophe Royoux, 67–76. New York: Sternberg, 2005.

———. "Technological World-Pictures." *Isis* 98, no. 1 (2007): 84–99.

———. "Interview with John Tresch." *Sonic Acts*, no. 20 (2015), accessed August 23, 2023, https://sonicacts.com/mobile/portal/interview-with-john-tresch-on-cosmograms.

Vandergeest, Peter, and Nancy Lee Peluso. "Territorialization and State Power in Thailand." *Theory and Society* 35 (1995): 385–426.

Vierenklee, Johann Ehrenfried. *Mathematische Anfangsgründe der Arithmetik und Geometrie: in so fern solche denjenigen, die sich dem höchstnöthigen Forstwesen auf eine vernünftige und gründliche Weise widmen wollen, zu wissen nöthig sind.* Leipzig: Weidmann, 1767.

Warde, Paul. *The Invention of Sustainability: Nature and Destiny, c.1500–1870.* New York: Cambridge University Press, 2018.

Weizman, Eyal. "Introduction." In *Forensis: The Architecture of Public Truth*, edited by Forensic Architecture, 9–32. Berlin: Sternberg Press, 2014.

3 Zooming Into Google Gaia Maps

From Globalization to Glocalization of Forests

Glocalization: Politics of Zoom and Global Synopticism

A basic feature of digital map systems like Google Earth is the ability to zoom. With the help of the zoom tool, conventionalized in the interface design by the icons of a looking glass and/or of a '+' or '-,' online maps like the platform Global Forest Watch (GFW) can be addressed. The technical term for this process, which is fully automated today, is downscaling. Prior to automation, the only way to achieve a similar zoom effect in maps—which were still made by hand at the time—was to laboriously produce more maps in other resolutions.

This method is also significant for climate visualizations. Downscaling climate information is broadly defined as 'transition across scales' to 'relate local- and regional-scale climate variables to the larger scale atmospheric forcing.'[1] These downscaled variables are always presented visually—as maps or elevations of specific locales—and may be accompanied by verbal narration or description of the climate 'scenarios' projected to obtain in that locale at a certain time under different models of CO2 forcing. Critical scholarship in environmental and cultural studies has established at least two political dilemmas generated in the process of downscaling: first, downscaled visualizations cancel out conflicting local data and, second, downscaled visualizations replicate the hegemonic dynamics of their global sources.

Downscaled climate visualizations are naturally resistant to locally situated views of climate that either contradict or are incompatible with quantitative models. While it is true that downscaled climate projections are 'ground-truthed' for a particular region with regional records of temperature, precipitation, etc., they cannot smoothly integrate the qualitative, spherical accounts of climate found in communities' stories, memories, or art. Furthermore, downscaled projections tend to cancel out local definitions of key parameters, such as vegetation type, in favor of globalized technical definitions. Examples of models of forestation—a key variable in climate change projections due to trees' significant carbon-storage capacity—will help illustrate the problem here.

We conceptualize the zoom tool with the term 'glocalization': the term refers to the coexistence, interplay, and connections of local and global

DOI: 10.4324/9781003376774-3

processes; Google Earth, in a continuous zoom, generates a completely scalable space that creates an aesthetically controllable, totalizing representation of the globe.[2] The zoom tool, as it is also central to the GFW platform, relates global and local/spherical scales in interactive climate visualizations and global forest maps not physically or literally but via a metaphor. A metaphor's unique function is to help a community understand a domain of knowledge or action that's unfamiliar to them by relating it to a familiar domain through specific features the two share. Metaphors help make abstract concepts concrete and actionable. At first, everyone recognizes a metaphor for what it is, but over time and habitual use, it becomes naturalized to the point where the community stops thinking of it as a tool that helps them interface with reality and starts thinking of it as reality itself.[3] So, for example, we're no longer aware that a 'window' was a particular, proprietary metaphor for a particular kind of graphical user interface on a personal computer; we just 'open' and 'close' windows on our Macs and PCs with no thought to who's paying the virtual heating bill.

'Zooming' is exactly this kind of naturalized metaphor. While multi-lens assemblies have been around since the 17th century, the descriptor 'zoom' was first used in a US patent application by Clile Allen in 1902—and, it was already metaphorical in that first usage, importing feelings of speed and travel from the domain of racing into the domain of optics. 'Zooming' quickly naturalized to become a standard term in photography, which it remained for much of the 20th century. But when it was applied to computer visualization (the *Oxford English Dictionary* reports the first usage in this sense from 1965 with increasing usage in the 1980s), the metaphorical sense of zoom was revived, importing into digital computing the analog norm of changing focal lengths from a fixed perspective, when what was really going on was the digital substitution on a fixed screen of a sequence of static 'snapshots' of an object, each displayed at a higher (or lower) resolution.

Zoom: Zoom is a metaphor developed in photography to describe the technical possibility of using media to change scales that exceed human vision. The zoom hides the incommensurable views, realities, and breaks that exist between the individual steps by representing them as continuous transitions. Upscaling and downscaling are part of the zoom concept.

Using this metaphor, climate service platforms like the World Bank Group's Climate Change Knowledge Portal (CCKP) have glocalized the 'god trick' into a personalized and detailed view for one's own use. For instance, a CCKP user can move the mouse over the world map to click on the region of Central America; the page refreshes, and Central America fills the map window; from there, the user can click on the country of Nicaragua, and the same procedure localizes the map window to the country's northern and southern borders. At the same time, a graph pops up to the right showing the average monthly temperature and rainfall for Nicaragua; users can manipulate the timescale of this graph to see roughly how these patterns have changed over the last century; and they can also zoom in to the timescale of each month to get more specific data. Such online platforms successfully realize Edward Tufte's visual information-seeking doctrine, which Ben Shneiderman later summarized with the formula 'overview first, zoom and filter, then details-on-demand.'[4]

Thus, interactive visual downscaling turns out not to be as automatic and transparent as the slider bar or zoom tool featured in many of these applications lead us to believe: rather, zooming is a metaphor that manufactures logical continuity out of what is in reality a diverse and sometimes incompatible set of views of climate composed at various scales. The seeming continuity of the visualization, the zoom, is an illusion composed of editing techniques such as fades, blends, and morphs. And Zachary Horton argues that this logical and technical process is also always political in the following way: 'contemporary scalar politics invests energy into singularities (individual heroes and villains, monuments, memes) and thus away from systems, while displacing undesirable consequences to nonvisible scales: the vast ocean, the atmosphere, the nano realm, the far future—comfortingly distant points on the scalar spectrum.'[5]

In terms of global forest visualization, then, what the zoom tool does politically is to impose global/transnational regimes of seeing, defining, and governing forests onto the local level. It represents deforestation in a particular tropical forest in the Global South, for instance, as a local instance of global climate change, thus authorizing well-resourced and connected companies in the Global North to interfere with that forest. In our previous work, we spent some time considering the Nicaraguan case introduced in Chapter One in which a REDD+ (Reducing Emissions from Deforestation and Forest Degradation) project in Nicaragua failed because Canadian forestry practices were being forced on local farmers according to neoliberal logics.[6] Now, we can go a step further and say that a politics of zoom is created when the zoom tool is used within a Google Gaia cosmology that defines 'local' as no more than a low-resolution piece of 'global,' a claim we will support below.

Edward Tufte's visual information-seeking doctrine is closely intertwined with the ideal of synoptic maps. Global synopticism is rooted in early geography and data visualization (governmental accounting and table work), of which Alexander von Humboldt's climate 'isothermal map' of the Northern Hemisphere (1817) is perhaps the most famous example.[7] Synopticism was valued as an epistemic ideal, the means by which Enlightenment scholars could transcend their physical limitations and observe all the laws that governed life on Earth: in other words, all-seeing was the first step to all-knowing. Since the 18th century, synopticism has promised to provide a deep understanding of superhuman-scale systems at a glance and has foregrounded the regularities within these systems to the detriment of differences. Finally, synopticism promoted efficiency by combining multiple datasets in a single viewframe, as epitomized by William Playfair's trade balance graphs, Joseph Priestley's historical charts (both 18th century), Joseph Minard's graph of Napoleon's disastrous Russian campaign (19th century), and Jacques Bertin's charts of visual variables (20th century).[8]

Synopticism: A scientific ideal, developed since the 18th century in early geography, public finance, social science, and data visualization, of providing a total overview and at a glance, based on the assumption that all-seeing is all-knowing.

It is hardly an exaggeration to say that 'climate change' as a concept would not exist without synoptic (global) views of weather and, later, concentrations of carbon dioxide and ozone. Climatography—the geographical part of meteorology—is the graphing of climate data onto world maps. By distributing geographical data across space, patterns and relations can be discovered. Nevertheless, previous research by critical humanities scholars has demonstrated that synoptic views of climate also participate in the formation of political inequalities.

This inequality can be most clearly seen in maps depicting social issues. Multiple scholars have made compelling cases for the ways in which synopticism and hegemony support each other to normalize populations. Michel Foucault, for example, found in John Graunt's synoptic tables of mortality in 17th-century London the seeds of the notion of 'population' and thus the 'birth of biopolitics,' in which rulers stopped governing individuals and started setting norms for group behavior. The synoptic shift led, in Foucault's view, to restrictive norms of gender, health, and sexuality that privileged white men without ever laying out an explicit moral case for this privilege: anything that was otherwise was simply disciplined out of power.[9] Donna Haraway narrowed Foucault's critique specifically to maps by theorizing the ways in which the 'god trick' of the synoptic empowers those who make the maps while disempowering viewers who have no authority or means to alter them.[10] Michel de Certeau termed this kind of disempowerment a 'strategy' of 'panoptic practice.'[11] Technical communication scholars have further established the disempowering effects on viewers of synoptic graphics such as global maps.[12]

In platforms like CCKP, the zoom metaphor exploits the history of camera development by taking on both a photographic (cartographic) and a cinematic aspect. The cartographic aspect has been sufficiently accounted for by the critical geography work reviewed earlier; the cinematic side of the metaphor, however, has received less analysis. The cinematic is invoked whenever synoptic views of the globe or climate are related to the detail (the singular data point or the closest gaze toward the ground) by simulation of a smooth and continuous process of zooming in on the Earth's surface. Within this conception, micro and macro levels are two ends of a unified continuum, conventionalized by the zoom slider tool of Google Maps.

Cinematic zooming may be most clearly exemplified by Charles and Ray Eames's famous short movie *Powers of Ten*, produced in 1968/1977 for an exhibition of the International Business Machines Corporation (IBM) (Figure 3.1). The Eameses' animated eight-minute moving picture smoothly closed the gaps between the distinct 'jumps' of the vertical space journey sketched out in the earlier book *Cosmic View: The Universe in 40 Jumps* by Kees Boeke (1957). *Powers of Ten* can be seen as a visual formula for Haraway's 'god trick.' The Eames movie reproduced on a cinematic level what already had become imaginable via globes, world maps, and later by space technology: a cosmic journey of a detached eye relating sky to ground along a vertical trajectory. The steady frame rate of the film minimized the 'jumps' that characterized the Boeke source, rendering a discontinuous process as a seemingly continuous one and thereby masking the not-insignificant problems of trying to reconcile images taken from different positions and at different scales.

But these problems persist. As Latour has pointed out, no human eye could maintain a steady view across the scales presented in *Powers of Ten*[13]; in fact,

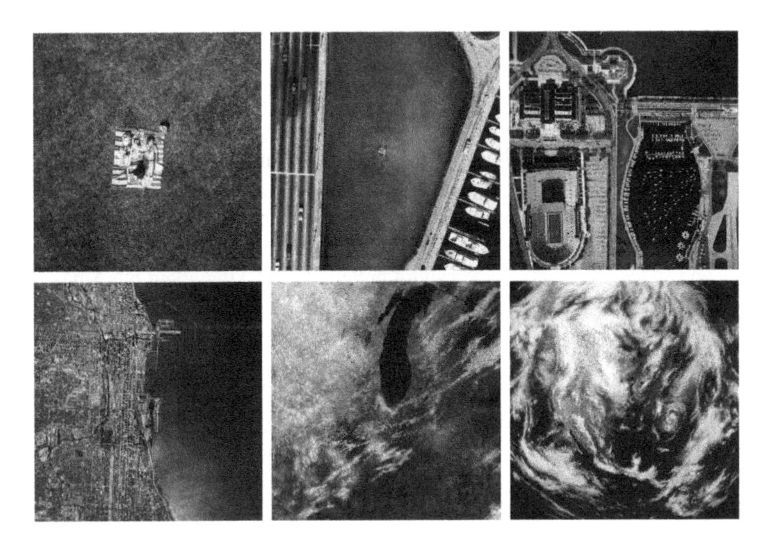

Figure 3.1 Six stills from *Powers of Ten* (1977), a film by Charles and Ray Eames.

Source: Figure adapted from Eames Foundation, details TBD.

some of the galaxies shown are not even visible to the human eye.[14] Even if we replaced the human eye with a camera, that camera could not record in a continuous fashion all the visualizations presented sequentially in the film; in fact, those visualizations come from dramatically different sources—optical and radio telescopes, film cameras, and animations. Thus, the cinematic move from the global to the local brings with it mutations in quality (i.e., the genre and source of images), not merely quantity (i.e., distance). The seeming continuity of the visualization, the 'zoom,' is an illusion composed of editing techniques such as fades, blends, and morphs. In this way, the zoom metaphor, a trope that aligns and compares disparate perspectives, turns into the zoom *synecdoche*, a trope that substitutes a part of something for the whole: under the force of the synecdoche of zoom, the global is viewed through the lens of the local.

> **Synecdoche:** A rhetorical figure in which a part of something is made to stand in for the whole, as happens when we refer to a perfumer as a 'nose.' Along with metaphor, metonymy, and irony, synecdoche is one of the four common tropes that Aristotle and ancient rhetoricians recognized as the dominant figures of artful expression.

This slippage from visual *comparison* via metaphor to visual *reduction* via synecdoche is consequential for the politics of interactive downscaled climate visualizations. As Latour puts the problem:

> [T]his illusion of unhindered movement limits reactions to the ecological crisis, since people think they can talk blandly about, for instance, 'everything,' or about the 'fate of the planet,' without realizing that what they call 'everything' generally tallies with some tiny model in a research bureau or lab. . . . Yet, it would be absurd to deny that differences in time and space are crucial. One cannot pretend that talking about the Amazonian Basin is the same thing as working on a ten-acre experimental station in the Jura.[15]

And yet that is precisely what happens with many zoom tool climate visualizations: the Amazonian forest is seen as part of Google Earth, and so by the logic of Gurevitch's 'Google Warming,' how Western scientists defined forests in Germany becomes the standard for how Amazonian forests are defined, regardless of the diversity of those ecological and political contexts. Examples of this kind of globalization can be located in REDD+ programs (in developing countries), whose investment capital frequently comes from nongovernmental organizations (NGOs) who market the carbon credits that farmers in the tropics, primarily Southeast Asia and Central and South America, produce by planting trees on their land. The NGOs and their corporate clients use global forestry standards and Google-based visualization tools to zoom in and check on the progress of trees and farmers. In spite of the good intentions of these programs, the zoom logics involved have supported some invasive neoliberal practices like the Nicaraguan case we introduced in Chapter One.[16]

Putting the politics of zoom back in contact with the history of forest mapping we reviewed in Chapter Two, we can ask, what now counts as a tree on digital maps produced by satellite imagery? In contrast to historical maps, the classification of trees and forests by satellite-based detection is intrinsically linked to the resolution of optical systems. The current highest resolution of satellite images of 5, 10, or 30 meters is the point where the zoom ends. There is no reality beyond these pixels; the limit of differentiation is reached. The resolution sets the condition to detect forests. The resolution, it is clear, is many times above the limit of human perception on the ground. Satellites are not eagle eyes gazing at the Earth but a view of a blurred world through thick glasses. Nevertheless, the synoptic view offers an additional value that transcends the visual awareness of a spectator on the ground; meanwhile, individual trees fall through this grid.

When forests were monitored globally in 2012 via remote sensing, tropical dry forests the size of the Amazon forest escaped the notice of satellite mapping because they fell through the cracks of the resolution of the remote-sensing algorithms.[17] The gap could only be closed with the help of more than 200 assistants from different research institutions who interpreted 200,000 individual VHR satellite images, which have a much higher resolution. After this process, estimates of global forest cover grew by nearly 10%. Here, local data integrated seamlessly and helpfully with global data. On the flip side, however, Paul Robbins found that an attempt to use local farmers' knowledge to ground-truth satellite estimates of forest cover in Rajasthan, India, failed—due to a mismatch between their definitions of 'forest' and official, governmental definitions inherited from transnational forestry agencies.[18] In short, farmers did not count as 'trees' the invasive, shrub-like species that the transnational scheme allowed to count as 'trees.' Thus, in the farmers' eyes, the official definition generated a grossly inflated estimate of reforestation, one that papered over the true environmental devastation of their homeland while allowing local governmental officials to secure transnational funding awarded to 'successful' national reforestation campaigns.

The labor connected to online forest maps is described in an essay by information scholar Cindy Lin with the example of West Kalimantan, Indonesia. In Indonesia, lower-class migrants were subcontracted as geospatial data technicians for Indonesia's National Mapping Agency 'to draw squares and rectangles around vast swaths of building and property' to differentiate primary forests and plantations and other types of land for the purpose of forest protection, after there had been conflicts in maps because of 'different forest classifications and mapping methodologies practiced by the two ministries.'[19] She quotes from an interview with one of the geospatial technicians: 'When the brown pixel changes into a green pixel, what you see is the edge of a forest and the beginnings of a house. You try your best to draw a square, but always, a little forest or house will escape,' and '. . . an extra pixel adds at least five meters to the total calculated area,' which means that drawing the right polygon is crucial. This example shows the complexity of the mapping process and how it is virtually impossible for an algorithm to get the right answer when processing satellite data looking for trees and forests.

If these are all the problems with zoom tool climate visualizations, what are the alternatives? How can we relate global and spherical views of climate to promote environmental justice without reducing climate to a synoptic, transnational perspective? Here, it is important to note that while global forest mapping platforms dominantly present Google Gaia cosmograms, they do create space for alternate views of forests, particularly when they support the non-reductive articulation of their global views with ground-up views and stories of forests and climate. We will see some of these opportunities emerge in our case study of GFW in Chapter Five. But researchers have also proposed abandoning global images altogether and in place of 'terrestrial' visions of Earth. 'Gaïa-graphy' of Critical Zones and Anti-Zoom,[20] for example, have been suggested as ways of generating maps that cannot be 'landed on' but can be lived in, thus motivating sustainable and just climate action around forests. It is these alternatives we will turn to in Chapter Four.

Notes

1 Bruce Hewitson and Robert G. Crane, "Climate Downscaling: Techniques and Application," *Climate Research* 7, no. 2 (1996): 85–95, 85.

2 Ursula K. Heise, *Sense of Place and Sense of Planet: The Environmental Imagination of the Global* (Oxford and New York: Oxford University Press, 2008); Ulrike Bergermann, ed., *Das Planetarische: Kultur, Technik, Medien Im Postglobalen Zeitalter* (München: Fink, 2010); Eva Horn and Hannes Bergthaller, *The Anthropocene: Key Issues for the Humanities* (London: Routledge, 2019).

3 Elisabeth El Refaie, "Metaphors We Discriminate By: Naturalized Themes in Austrian Newspaper Articles About Asylum Seekers," *Journal of Sociolinguistics* 5, no. 3 (2001): 352–71; George Lakoff and Mark Johnson, *Metaphors We Live By* (Chicago: University of Chicago Press, 1980), 245–7.

4 Ben Shneiderman, "The Eyes Have It: A Task by Data Type Taxonomy for Information Visualizations," *Paper Presented at the Proceedings 1996 IEEE Symposium on Visual Languages*, September 3–6, 1996, 337, https://ieeexplore.ieee.org/document/545307; Edward R. Tufte, *The Visual Display of Quantitative Information*, Vol. 2 (Cheshire, CT: Graphics Press, 2001).

5 Zachary Horton, *The Cosmic Zoom: Scale, Knowledge, and Mediation* (Chicago, IL: University of Chicago Press, 2021), 6.

6 cf. Birgit Müller, "Farmers, Development, and the Temptation of Nitrogen: Controversies About Sustainable Farming in Nicaragua," *RCC Perspectives* 5 (2012): 23–30.

7 Birgit Schneider, "Rendering Visible the Climate. Humboldt's 1817 Climate Zone Map," *MLN—Modern Language Notes*, Special Issue In/Visible 137, no. 3 (2022): 545–65. Birgit Schneider, *Klimabilder. Eine Genealogie Globaler Bildpolitiken Von Klima Und Klimawandel* (Berlin: Matthes & Seitz, 2018).
8 For these examples and others, see Tufte, *Visual*.
9 Michel Foucault, "Security, Territory and Population," in *Michel Foucault, Lectures at the College De France* (Basingstoke: Palgrave Macmillan, 2007), 103.
10 Donna Jeanne Haraway, *Modest-Witness@Second-Millennium.Femaleman-Meets-Oncomouse : Feminism and Technoscience* (New York: Routledge, 1997), 136.
11 Synopticism is a kind of panopticism. Michel de Certeau, *The Practice of Everyday Life* (Berkeley: University of California Press, 2011).
12 Ben F. Barton and Marthalee S. Barton, "Modes of Power in Technical and Professional Visuals," *Journal of Business and Technical Communication* 7, no. 1 (1993): 138–62, Lee E. Brasseur, *Visualizing Technical Information: A Cultural Critique*. Baywood's Technical Communications Series (Amityville, NY: Baywood Pub, 2003); Gunther Kress and Theo van Leeuwen, *Reading Images: The Grammar of Visual Design* (New York: Routledge, 1996); Sam Dragga and Dan Voss, "Cruel Pies: The Inhumanity of Technical Illustrations," *Technical Communication* 48 (2001): 265–274.
13 Bruno Latour, "Anti-Zoom," in *Contact, catalogue de l'exposition d'Olafur Eliasson* (Paris: Fondation Vuitton, 2014).
14 Lynda Walsh and Lawrence J. Prelli, "Getting Down in the Weeds to Get a God's-Eye View: The Synoptic Topology of Early American Ecology," in *Topologies as Techniques for a Post-Critical Rhetoric*, ed. Lynda Walsh and Casey Boyle (New York: Palgrave Macmillan, 2017), 197–218.
15 Latour, "Anti-Zoom," 121.
16 Michael K. McCall, "Beyond 'Landscape' in REDD+: The Imperative for 'Territory'," *World Development* 85 (September 1, 2016): 58–72, https://doi.org/10.1016/j.worlddev.2016.05.001.
17 Betsabé de la Barreda-Bautista, "Tropical Dry Forests in the Global Picture: The Challenge of Remote Sensing-Based Change Detection in Tropical Dry Environments," in *Planet Earth 2011—Global Warming Challenges and Opportunities for Policy and Practice*, ed. Elias Carayanni (InTech: Open Access Publisher), 231–256, DOI: 10.5772/24283.
18 Paul Robbins, "Fixed Categories in a Portable Landscape: The Causes and Consequences of Land-Cover Categorization," *Environment and Planning A* 33, no. 1 (2001): 161–79.
19 Cindy Lin, "How to Make a Forest," *E-Flux Architecture*, 2020, www.e-flux.com/architecture/at-the-border/325757/how-to-make-a-forest/
20 Latour, "Anti-Zoom"; Alexandra Arènes, Bruno Latour, and Jérôme Gaillardet, "Giving Depth to the Surface: An Exercise in the Gaia-Graphy of Critical Zones," *The Anthropocene Review* 5, no. 2 (2018): 120–35.

Bibliography

Barreda-Bautista, Betsabé de la. "Tropical Dry Forests in the Global Picture: The Challenge of Remote Sensing-Based Change Detection in Tropical Dry Environments." In *Planet Earth 2011—Global Warming Challenges and Opportunities for Policy and Practice*, edited by Elias Carayanni, 231–56. Tech: Open Access Publisher, 2011, DOI: 10.5772/24283.

Barton, Ben F., and Marthalee S. Barton. "Modes of Power in Technical and Professional Visuals." *Journal of Business and Technical Communication* 7, no. 1

(January 1, 1993): 138–62, https://doi.org/10.1177/1050651993007001007, http://jbt.sagepub.com/content/7/1/138.abstract.

Bergermann, Ulrike, ed. *Das Planetarische: Kultur, Technik, Medien Im Postglobalen Zeitalter*. München: Fink, 2010.

Brasseur, Lee E. *Visualizing Technical Information: A Cultural Critique*. Baywood's Technical Communications Series. Amityville, NY: Baywood Pub, 2003.

Certeau, Michel de. *The Practice of Everyday Life*. Berkeley: University of California Press, 2011.

Dragga, Sam, and Dan Voss. "Cruel Pies: The Inhumanity of Technical Illustrations." *Technical Communication* 48 (2001): 265–74.

El Refaie, Elisabeth. "Metaphors We Discriminate By: Naturalized Themes in Austrian Newspaper Articles About Asylum Seekers." *Journal of Sociolinguistics* 5, no. 3 (2001): 352–71.

Foucault, Michel. *Security, Territory and Population*. Michel Foucault, Lectures at the College De France. Basingstoke: Palgrave Macmillan, 2007.

Haraway, Donna J. *Modest_Witness@Second_Millennium.Femaleman(C)_Meets_Oncomousetm: Feminism and Technoscience*. New York: Routledge, 1997.

Heise, Ursula K. *Sense of Place and Sense of Planet: The Environmental Imagination of the Global*. Oxford and New York: Oxford University Press, 2008.

Hewitson, Bruce, and Robert G. Crane. "Climate Downscaling: Techniques and Application." *Climate Research* 7, no. 2 (1996): 85–95.

Horn, Eva, and Hannes Bergthaller. *The Anthropocene: Key Issues for the Humanities*. London: Routledge, 2019.

Kress, Gunther, and Theo van Leeuwen. *Reading Images: The Grammar of Visual Design*. New York: Routledge, 1996.

Lakoff, George, and Mark Johnson. *Metaphors We Live By*. Chicago, IL: University of Chicago, 1980.

Latour, Bruno. "Anti-Zoom." Contact, catalogue de l'exposition d'Olafur Eliasson, Paris, Fondation Vuitton, 2014.

Lin, Cindy. "How to Make a Forest." *E-Flux Architecture*, 2020, www.e-flux.com/architecture/at-the-border/325757/how-to-make-a-forest/

McCall, Michael K. "Beyond 'Landscape' in REDD+: The Imperative for 'Territory'." *World Development* 85 (September 1, 2016): 58–72, https://doi.org/10.1016/j.worlddev.2016.05.001.

Robbins, Paul. "Fixed Categories in a Portable Landscape: The Causes and Consequences of Land-Cover Categorization." *Environment and Planning A* 33, no. 1 (2001): 161–79, https://doi.org/10.1068/a3379.

Schneider, Birgit. "Rendering Visible the Climate. Humboldt's 1817 Climate Zone Map." *MLN—Modern Language Notes*, Special Issue In/Visible 137, no. 3 (2022): 545–65.

Shneiderman, Ben. "The Eyes Have It: A Task by Data Type Taxonomy for Information Visualizations." *Paper Presented at the Proceedings 1996 IEEE Symposium on Visual Languages*, September 3–6, 1996, https://ieeexplore.ieee.org/document/545307.

Tufte, Edward R. *The Visual Display of Quantitative Information*, Vol. 2. Cheshire, CT: Graphics Press, 2021.

Walsh, Lynda, and Lawrence J. Prelli. "Getting Down in the Weeds to Get a God's-Eye View: The Synoptic Topology of Early American Ecology." In *Topologies as Techniques for a Post-Critical Rhetoric*, edited by Lynda Walsh and Casey Boyle, 197–218. New York: Palgrave Macmillan, 2017.

4 Forests as Stories

Storyworld Networks as Alternatives to Google Gaia

As is hopefully clear by this point in the book, blue marble, red marble, and green marble data visualizations of the Earth are all cosmograms: that is, they're holographs that depict in one image an entire cosmology (worldview). We have named the cosmology represented by green marble cosmograms Google Gaia, and we have further argued that they work through a Plantationocene' lens to grid the Earth's forests so they can be capitalized, budgeted, and banked by industrialized nations as offsets for the global warming their actions produce. The 'plan' part of the plantation usually takes place at computer terminals in the Global North, thousands of miles away from the actual forests being monitored and managed in the Global South.

If the footnotes from Chapters Two and Three are any indication, we are certainly not the first authors to recognize that these visual dislocations, discontinuities, and dissociations participate in efforts to impose colonial world views on the colonized. Multiple contributors to *Feral Atlas*, *Arts of Living on a Damaged Planet*, and *Critical Zones* address this dynamic, which we call *cosmological imperialism*.

> **Cosmological imperialism:** The use of images of 'the world' as part of a neo/colonial propaganda campaign aimed at imprinting the colonizer's ideology (religion, political philosophy, economy, etc.) on the colonized, thereby making them a colony in part or whole.

Our question now is, 'Where to from here?'[1] If cosmological imperialism is the problem in global forest visualization, what is/are the solution(s)? This question has also been raised in some form by Anna Tsing, Donna Haraway, Robin Wall Kimmerer, and other scholars working on the intersection of science, the humanities, and environmental justice (EJ).[2] And, they have identified a range of alternative visualizations, many of them Indigenous and decolonial in nature. In this chapter, we will review these alternatives to Google Gaia, making sense of them with the help of anthropologist Tim

DOI: 10.4324/9781003376774-4

Ingold's globes/spheres distinction. Then, we will wrestle with the objection put forward by some creators of these alternatives—namely, that true EJ demands doing away with Google Gaia altogether. While acknowledging the wisdom of this position, we also know from our work that global forest visualization platforms like GFW have proven useful to some decolonial and Indigenous activists. Accordingly, we will end this chapter by suggesting a compromise, a hybrid, between pro- and anti-Google Gaia positions. This hybrid position, which we call storyworld networking, lets climate justice advocates continue to use global forest visualization platforms in concert with local/Indigenous cosmograms of forests.

Globes and Spheres

In considering the same history of colonial maps that we reviewed in Chapter Two, Tim Ingold identified two very different ways of mapping the Earth and its environment, typified largely (but not entirely) by differences between colonial and Indigenous cosmologies. He determined that colonial cosmologies tend to generate *global* views of the environment, which he characterizes as top-down, outside-in, hard, and visual, among other features. In contrast, Indigenous environmental views tend to be *spherical*—bottom-up, inside-out, soft, acoustic, and experiential (Figure 4.1).[3]

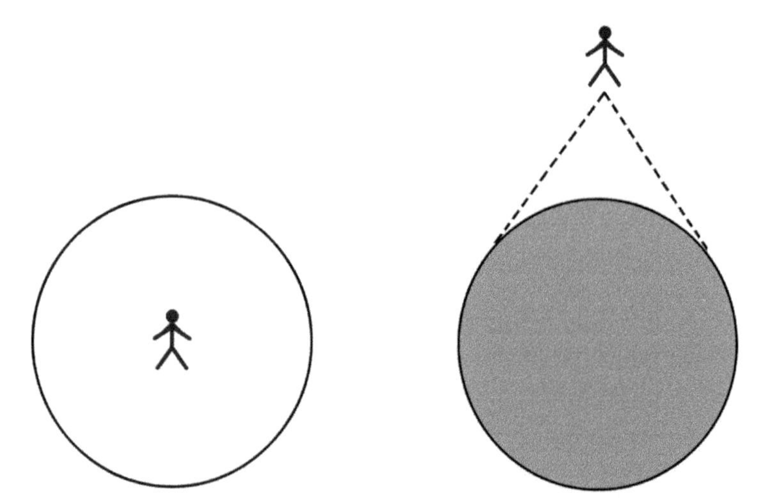

Figure 4.1 Illustration of spherical view (a) versus global view (b) of environment and climate.

Source: Figure adapted from Ingold, Tim. 'Globes and Spheres: The Topology of Environmentalism.' In Environmentalism: The View from Anthropology, edited by Kay Milton, 31–42. London: Routledge, 1993.

By drawing this distinction, Ingold importantly recasts the traditional opposition between global/local; for him, both those terms still participate in the global paradigm, one that just sees the Earth in terms of its surface. A spherical paradigm instead places the viewer *inside* the environment: when the viewer moves, so does the environment; when the viewer changes, so does the environment. This is how people can effect environmental change in Ingold's theory. In other words, where globes form the scene of observation, spheres form the scene of action.

With this framework in place, we can observe that the alternatives to Google Gaia that have been articulated in *Critical Zones* and elsewhere tend to engage either global or spherical tactics: in other words, they either disrupt global views of the Earth by distorting scales, coding features, or other *counter-mapping* strategies; or, they replace globes with spheres by creating *dwellings*. Of course, they can and do combine these techniques, but for the sake of clarity we will treat examples under these two headings.

Counter-maps

Nancy Peluso was the first to define 'counter-mapping' as a practice in her 1992 study of how local Indonesian populations were resisting palm oil plantations by creating new maps of their territories that contested the boundaries drawn on state and corporate maps. Peluso writes that in making the counter-maps, the Kalimantan activists sought to 'appropriate the state's *techniques* and *manner of representation* to bolster the legitimacy of "customary" claims to resources.'[4] The key point in this definition is that counter-maps employ the maps they wish to challenge as a visual foundation so that their counter-claims to sovereignty will be visible to legal authorities.

Counter-mapping strategies can thus be arranged roughly along a continuum based on how obvious the base colonial map remains in the final accounting. At the minimum deviation end of the spectrum would be the Kalimantan maps, which drew key boundaries around tribal lands while leaving the old boundaries mostly intact. Other examples include citizen hazard mapping projects, such as the SafeCast mapping project following the Fukushima tsunami in which citizens took their own radiation readings with handheld Geiger counters and pinned them to maps that disagreed sharply in some instances with official governmental radiation estimates for those areas[5]; and the Pepys Estate noise maps populated with decibel readings recorded by residents seeking to get the London City Council to pass ordinances against industrial sound waste in their neighborhood.[6] Many decolonial projects use this strategy as well, such as Mariel Rodriguez's 'Flowers of Evil I' installation @ Artistic Bokeh in Vienna, which penciled in coca trading routes over an old Austrian colonial map and linked key ports to a string diagram of Viennese cocaine advertisements and photographs of its paraphernalia and use.[7]

At the other end of the spectrum, we find acoustic maps such as Sacha Taki birdsong 'soundscape' created by the Ecuadorian Pueblo Ancestral Kichwa Kawsak Sacha[8] and the 'living maps' created by the Marind of West Papua, who were contesting the destruction of their forests by palm oil plantations. Sophie Chao, who worked with the Marind to produce the maps, did begin with colonial basemaps of the area, but as the Marind overmarked them with the 'lifeways' of the nonhuman beings that were important to their survival, 'Concession and administrative boundaries disappeared under a dense meshwork of intertwined, interconnected and multitemporal existences.'[9] Chao transferred this meshwork into a digital platform that could also support the integration of audio files with birdsongs and other natural sounds at particular places on the map. But she discovered that 'Marind maps themselves keep morphing and never sit still, in the image of the multispecies world itself.'[10] These dynamics created problems for the intelligibility of the maps to government agents; in fact, the project was never completed. Chao concluded that for the Marind, counter-mapping was in large part an opting out from cosmological imperialism.

Another example of a disruptive or resistant counter-map can be located in the work of Hélio Melo, a self-taught artist who worked his whole life as a rubber tapper in Brazilian Amazon. For example, in a pictorial map that borrowed its form from an enormous rubber tree, Melo marks out the laborious path he had to follow through the jungle every day in order to tap enough rubber to pay his debts to the company he works for—a company that failed to make rubber trees grow successfully in plantation in the Americas and so resorted to wage slavery to turn a profit in the global rubber market. Set in its context, Melo's beautiful map includes a layer of quiet protest.[11]

Also at this end of the counter-mapping spectrum, we encounter projects such as Mixpantli, an art installation at the Los Angeles County Museum of Art in which Mariana Castillo Deball and Sandy Rodriguez have both reworked the oldest synoptic map of Mexico City. Commissioned by the Spanish but drawn by an indigenous cartographer, Tenochtitlan (1521) was already something of a counter-map itself, as it bucked European cadastral conventions in favor of larger-than-life images of Indigenous people going about their business in the city and surrounding landscapes. Deball and Rodriguez's interventions further disrupt the colonial mapping tradition by fragmenting the original map in rubbings and then blowing them up so large that visitors must walk on the map and cannot achieve a synoptic view of it; in addition, the artists re-illustrate sections of the map with contemporary visual stories about war, femicide, immigration, and other issues displayed on the walls of the exhibition space.[12]

In between these two extremes of counter-mapping, we find projects such as the *Feral Atlas*, which mixes temporal and spatial scales to create multimedia, multi-layered, multi-species story-maps of capitalistic colonization in the Americas. Editors Tsing et al. recruited scientists, Indigenous artists, architects, and other specialists to create these collaborative visualizations.[13] For instance, contributors Aït-Touati, Arènes, and Grégoire generated visual story-maps of normally invisible soils and the effects of capitalistic agriculture on them over time.[14] Another example of counter-mapping can be located in the Ciclos anuais calendar. This project, started after a visit by Indigenous Amazonians to the Berlin Ethnological Museum's trove of artifacts taken from their territories by Germans at the turn of the 20th century, assumes the form of a traditional Western zodiacal sky-wheel. However, it segments that wheel according to the traditional constellations of the Indigenous riparians along the Río Tiquié, populating it with rings (and layers) of information kept by local custodians about seasonal rainfall, animal life, and plant harvests.[15]

As we head back toward the more traditional end of the counter-mapping spectrum, we find several projects by Australian Aboriginal activists. The Saltwater Collection was a series of 50 bark paintings submitted in lieu of traditional written documentation to a 2008 Australian lawsuit over commercial fishing in traditional Yolŋu tidal zones in Blue Mud Bay. Although the area of dispute was clearly delineated in accompanying satellite maps, the Saltwater Collection itself did not follow these contours. Rather, the bark strips depicted tidal patterns, sea life, fishing equipment, and boats that together comprised a record of almost 50,000 years of use of the tidal zones; the maps succeeded in establishing the legal priority of the Yolŋu in these areas.[16]

The maps produced by the Paruku Project were more recognizable to outsiders, or at least, they began that way—with base geo-coordinate maps of the Paruku (Lake Gregory) region in Western Australia, over which Walmajarri custodians painted layer after layer of acrylics to tell stories of 'Country' or the hybrid social/natural/religious history of their territory.[17] Making these maps was the primary means of communication between the custodians and the Euro-American artists, archaeologists, biologists, and fire ecologists they had invited to work with them on key problems they were facing (e.g., settlement disputes, fish parasites, and drought). The finished maps varied quite a bit in terms of their final fidelity to the base topographical maps—some still clearly delineated the lake and surrounding territory, while others were much more like the Blue Mud Bay maps and were only intelligible as such to Walmajarri custodians. See Figure 4.2 for an example of such a counter-map.

In this brief survey of counter-mapping traditions, we can see Peluso's logic operating of 'appropriating the techniques' of cosmological imperialism. Not so for the next set of disruptive solutions, which take to heart Audre Lorde's decolonial warning that 'the master's tools will never dismantle the master's house.'[18]

Dwellings

The inside-out logic of dwellings as anti-synoptic visualizations of environment and climate owes a great debt once again to Indigenous traditions of

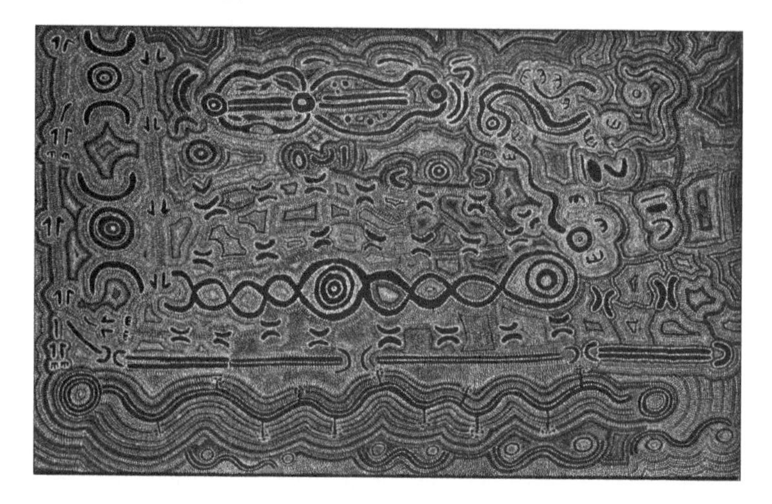

Figure 4.2 Aboriginal Religious Art (6854184762) from the collection of the St. Mungo Museum of Religious Life & Art, Glasgow. This is an Australian Aboriginal map of the country, a synoptic view that contains not only topographic locations such as waterholes (circles) and villages (horseshoe shapes) but also layers of kinship and historical information.

Source: Figure adapted from Wikimedia Commons.

storytelling—but also to feminist theory, which at least since Lorde's galvanizing speech at the 1984 NYU Institute for the Humanities Forest utopias has cast doubts on any liberatory movement that comes from within the establishment. The argument is simple and can be summed up by Einstein's (apocryphal) saying that we cannot hope to solve a problem using the same kind of thinking that created it in the first place. Drawing this logic into the arena of global forest visualization, we could put the argument like this: no matter how you twist, turn, or cover up a geo-coordinate basemap of a forest, its structural logics of oppression will seep through and stain your project—whether by constraining funding, creating translation difficulties, or simply failing to support any action that appears to threaten the establishment.

One solution is simply to refuse to visualize at all a radical disruption of the 'endless cyclopean war story from above' that Donna Haraway identifies with global patriarchy.[19] We saw the seeds of this approach in the insistence of the Marind on moving away from the visual in their counter-mapping exercises and their ultimate refusal to participate in the assemblage of a final visual map. However, even the most staunch advocates of feminist storytelling include visuals: Haraway's book covers, for instance, always feature important pieces of feminist visual art. When the visual mode is engaged by these disruptive activists, it tends toward Ingold's 'spherical' category—building rich visualizations of grounded, lived experiences of forests and environments from the bottom up and the inside out. These lifeworld bubbles radiate out from the human or nonhuman being at the center and focus on the things that being needs to flourish. Thus, we refer to this mode of disruptive visualization as dwelling.

The oldest examples of these kinds of dwellings, with respect to forests at least, can be located in tree ceremonies, which take place in one form or another in almost every part of the world. In West Africa, designated fetish trees outside villages are adorned with cotton balls signifying wishes and prayers of petitioners. In North America, various plains peoples plant a tree trunk in the ground, usually from a white birch, and perform sundance rituals around it. In many European traditions, dancers weave intricate ribbon designs around a central 'tree' or pole during maypole rituals. All of these ceremonies are cosmograms in that they are designed to balance or order the community's world from the inside out; yet, they are ephemeral and impossible to comprehend when viewed top-down from a synoptic global perspective.

The other type of forest dwelling that most readers are probably familiar with is the utopian forest village. Always presented as an alternative to or escape from the 'normal world,' these villages feature houses built into or between trees that cannot be accessed without the good will of the inhabitants. Some famous examples include the Ewok Village from *Star Wars: Return of the Jedi* and the Tipani Hometree from the movie *Avatar*. But, these types of dwellings are also constructed by activist defenders of redwood and other old-growth forests, who climb up into the canopies of key trees and create short- or long-term homes there, as Nick and Olivia do in Richard Powers's *Overstory*. This strategy melds the sphere of the tree's dwelling with the sphere of the humans, turning any attack on the tree into an attack on a human. Utopian tree villages are paradoxically highly defensible while also being extraordinarily vulnerable to damage to their substrate trees.

If putting houses in trees is one way of creating a forest dwelling, another is putting trees in houses. Traditional East Asian homes—working from ancient Confucian, Shinto, and other principles—frequently incorporate one or more trees into their central atrium gardens. The Hundertwasser House in Vienna included 'tree tenants,' for which growth corridors were built up and through the surrounding apartments (Figure 4.3). The Bosco Verticale project in Milan incorporates roughly 800 trees into its balconies and atria, which, along with an additional 19,000+ shrubs and plants, is projected by its designers to convert over 44,000 tons of carbon to oxygen and plant growth—although this outcome has been deemed unlikely, rendering the Bosco Verticale likely a purely symbolic healing of the ills of urban-industrial living.[20]

Whether treehouses or trees-in-houses, these cosmograms of dwelling resist synopticism by telling visual stories of forests whose reference point is not a satellite in space, but a squirrel, a mushroom, or a human on the ground (or in the branches). These are stories with first-person, not omniscient narrators; as a result, they necessarily focus on what is needed to sustain life.

Figure 4.3 The Hundertwasser House, an experimental apartment house built by Viennese architect Friedrich Stowasser in 1985 that incorporates 250 'tree tenants' in its design.

Source: Figure adapted from author photo.

Arguments for and Against Google Gaia in Climate Justice Activism

Several of the disruptive visualization frameworks above come with calls to get rid of global visualizations altogether. This is the 'strong' decolonization argument—the rationale being that the 'master' and all his tools must be thrown out of a territory in order for its Indigenous inhabitants to flourish in the way they did before colonization.

For communities who conclude they must reject any form of global visualization of their territories in order to accomplish decolonization, this is often part of a larger movement toward *data sovereignty*. The 1983 Pine Gap protests exemplify this stance, as Aboriginal and non-Aboriginal Australian women gathered at Australia's largest satellite surveillance facility to protest the intersection of surveillance and the oppression of Indigenous peoples, among other toxic dynamics.[21]

Data sovereignty: Generally speaking, data sovereignty is the right of any nation to govern data about its territory, people, and culture. In Indigenous contexts, data sovereignty was one of the key tenets of the 2007 United Nations Declaration on the Rights of Indigenous People (UNDRIP); in application, it is frequently assessed using a CARE metric (Collective Benefit, Authority to Control, Responsibility, and Ethics). The 2016 Te Mana Rauranga charter, ratified by the Māori nation, is considered a germinal document for Indigenous data sovereignty movements.

But the rejection of global surveillance is not a solution that fits all communities. Those invested in counter-mapping, for instance, need to register their maps within some global synoptic framework of measurement if they want to make their critique of the colonial situation legible to outside allies and authorities. We witnessed this strategy in both the Blue Mud Bay and Paruku cases described earlier.

So, while respecting all of the criticisms raised against Google Gaia cosmograms such as GFW, we would nevertheless like to suggest at least one way in which local and Indigenous activists can articulate those frameworks with their own local visualizations in a way that expands rather than reduces views of the forest and opportunities for climate action. We call this new framework *storyworld networking*, a kind of visual storytelling aimed at making worlds we want to live in.

Storyworld Networking

We have found a path toward integrating spherical and global visualizations of forests—without flattening one into the other—by articulating Tresch's cosmographic theory to two other concepts: Tim Ingold's globes/spheres distinction and David Turner's Anthropocene narrative. We call this framework storyworld networking.

Storyworld networking: A visual narrative network of global images of the Earth, such as the green marble, that serve as key political nodes in both time and space. In time, they catalyze a transition from an old worldview to a new one; in space, they articulate communities with each other. These articulations in both dimensions generate new affordances for political action while cutting off others.

Ecosystem scientist David Turner has written a monograph on the green marble in which he ties the image specifically into an 'Anthropocene narrative' that transitions from one 'sphere' or cosmogram into another via a period of rupture and reorganization (Table 4.1):

Table 4.1 Turner's Anthropocene narrative framework.

Sphere	*Epoch*
Geosphere/biosphere	'Gaian self-organization' in pre-human times (p. 13)
Great separation	Colonial period/industrial revolution: global maps appear
Technosphere	Humanity develops enough technological prowess to insulate itself from the Gaian biosphere
Great acceleration	Post WWII
Technosphere 2.0	Technosphere embraces biosphere
Great transition	1970s: anxieties about impacts of technosphere on biosphere: blue marble appears
Technobiosphere	Current epoch: green marble appears

In the future of this Anthropocene narrative, Turner predicts a period of equilibration, when 'humanity learns to self-regulate and manage the Earth system.'[22] Turner expresses hope that green marble will prove useful for ushering us into equilibration (this concept is related to the concepts of the noosphere and the technosphere.[23]

Here it is worth remembering Ingold's distinction between spherical and global ways of imaging the Earth. Whereas Turner makes no such distinction between top-down images of the Earth like the green marble and inside-out conceptions of Anthropocene life like the technobiosphere, we can nevertheless observe that global images—for example, the first global maps, the blue marble, and the green marble—emerge precisely at transitions between one sphere and the next in Turner's narrative scheme. Therefore, we can posit that global images serve as turning points in the plot of the Anthropocene narrative, by drawing us out of our current spherical worldview into an outside, global perspective, which then prompts us to reinvent, or 'redescribe,' in Tresch's words, our former worldview. In fact, Tresch goes on to argue that a cosmogram like the green marble can serve as just such a narrative catalyst:

> In this sense, cosmograms have a relation to time like that of the rites of passage that all societies have: the liminal time in which ordinary relations are suspended, in which there's often a symbolic recreation of the world and of society, at the same time as the formation of a community outside of ordinary social structures. After the ritual sequence, the participants come back to a transformed world, with the structures redefined, the cosmos remade: the space of possibilities is closed up again. *Cosmograms often guide this recreation and restabilization of the world.*[24]

So, in reconciling Tresch's theory to Ingold's and Turner's, we arrive at *storyworld networking*. Here's how it works: the image of blue marble appeared at the very moment when Euro-American scientists and activists started to notice 'cracks'[25] in the technosphere—that is, when they realized that humanity's enormous technical leverage was starting to bend and break the biosphere. At that turning point, the blue marble image brought the old Gaian biosphere back into view; that image jolted viewers out of the technosphere and ushered them into the technobiosphere. This chain of interaction forms a *story* driven by these particular images of the *world* distributed across both time and space: this story both supports new kinds of political action (e.g., climate change legislation) while cutting off older/other kinds of action (e.g., praying to the gods to stop climate change, as the technobiosphere presupposes that natural problems have natural causes and technical solutions).

While we arrived at our term storyworld simply by translating the Greek roots of 'cosmology' in a way that draws out the ancient sense of 'story' embedded in 'logos,' it's important to note we found a pre-existing, resonant concept of the storyworld in narrative theory and criticism. In that field, the term was first used by David Herman to define a narrative as not merely lines on a page or vibrations in a hearer's ear but as a world that the teller builds with their words in order to persuade the reader/hearer to inhabit it with them, even if just for a while.[26] Erin James, who has found storyworlds a natural fit for the kinds of Anthropocene narratives she studies as an ecocritic, sums up the narrative as 'worldbuilding for some purpose.'[27] While these theorists usually work with written stories, Herman at least has expanded the storyworld concept to graphic novels.[28] We believe we, too, are dealing with 'worldbuilding for a purpose' in the ways in which designers and users work with global forest visualization platforms like GFW. We understand these platforms not just as visualization tools but as evolutionary catalysts that articulate worldviews (cosmologies) across temporal and spatial scales into a coherent narrative about how the world works and what we should do about it.

In this way, our work aligns with EJ scholars who propose narrative networking as a better way to trace paths from particular visualizations of climate change to political action.[29] Finally, storyworld networking as a methodological approach equips us to analyze the political impacts of GFW—to observe (a) how its global-scale images (cosmograms) articulate worldviews (cosmologies) and (b) how its cosmograms articulate communities, both human and nonhuman.[30] Accordingly, in the next chapter, we examine the ways that GFW articulates globes and spheres, images and stories in order to determine what kind of storyworld network the platform creates for its users and what the politics of that story may be.

Notes

1 Nedra Reynolds, *Geographies of Writing: Inhabiting Places and Encountering Difference* (Carbondale, IL: University of Southern Illinois Press, 2004), 82.
2 Donna J. Haraway, *Staying with the Trouble: Making Kin in the Chthulucene* (Durham, NC: Duke University Press, 2016), 49; Robin Wall Kimmerer, *Braiding Sweetgrass: Indigenous Wisdom, Scientific Knowledge and the Teachings of Plants* (Minneapolis, MN: Milkweed Editions, 2013), 328; Anna Lowenhaupt Tsing, *The Mushroom at the End of the World: On the Possibility of Life in Capitalist Ruins* (Princeton, NJ: Princeton University Press, 2015), 190.
3 Tim Ingold, "Globes and Spheres: The Topology of Environmentalism," in *Environmentalism: The View from Anthropology*, ed. Kay Milton (London: Routledge, 1993).
4 Nancy Lee Peluso, "Whose Woods Are These? Counter-Mapping Forest Territories in Kalimantan, Indonesia," *Antipode* 27, no. 4 (1995): 384.
5 "Safecast Map," 2011, accessed August 11, 2023, https://map.safecast.org.
6 James Wynn, *Citizen Science in the Digital Age: Rhetoric, Science, and Public Engagement* (Birmingham: University of Alabama Press, 2017), 128–62.
7 Mariel Rodriguez, "Flowers of Evil/Blümen des Übels," 2017, https://marielrodriguez.hotglue.me/?flowers.
8 "Sacha Taki," *Pueblo Ancestral Kichwa Kawsak Sacha (PAKKS)*, 2019, accessed July 26, 2023, https://pakks.org.ec/que-hacemos/.
9 Sophie Chao, "'There Are No Straight Lines in Nature': Making Living Maps in West Papua," *Anthropology Now* 9, no. 1 (2017): 24.
10 Ibid., 17.
11 José Roca, "The Whole of Acre in a Single Tree: Brazilian History Through the Work of the Self-Taught Artist Hélio Melo, A Rubber Tapper from the Amazonas," 2008, accessed August 11, 2023, https://universes.art/en/magazine/articles/2008/helio-melo.
12 Sandy Rodriguez and Mariana Castillo Deball, "Mixpantli: Contemporary Echoes," December 12, 2021–June 12, 2022, accessed August 11, 2023, www.lacma.org/art/exhibition/mixpantli-contemporary-echoes.
13 Anna L. Tsing, et al., *Feral Atlas: The More-than-Human Anthropocene* (Stanford: Stanford University Press, 2020).
14 "Terra Forma, Mapping Ruined Soils," 2020, accessed July 26, 2023, https://feralatlas.supdigital.org/?cd=true&rr=true&cdex=true&text=terra-forma-mapping-ruined-soils&ttype=essay.
15 "O Ciclo Annual no Río Tiquié," 2019, accessed July 25, 2023, www.mikkogaestel.com/stage/o-ciclo-anual-no-rio-tiquie/.
16 "Gapu-Monuk Saltwater," 2018, accessed August 11, 2023, www.sea.museum/saltwater.
17 Steve Morton, et al., eds., *Desert Lake: Art, Science and Stories from Paruku* (Collingwood: CSIRO, 2013).
18 Audre Lorde, "The Master's Tools Will Never Dismantle the Master's House," in *Sister Outsider: Essays and Speeches* (Berkeley, CA: Crossing Press, 1984, 2007), 112.
19 Donna Haraway, "Carrier Bags for Critical Zones," in *Critical Zones: The Science and Politics of Landing on Earth*, ed. Bruno Latour and Peter Weibel (Cambridge, MA: MIT Press, 2020), 440.

20 Birgit Schneider, "Entangled Trees and Arboreal Networks of Sensitive Environments," *ZMK Zeitschrift für Medien-und Kulturforschung* 9, no. 1 (2018).
21 Felicity Ruby, "Minding the Gap," *Arena Magazine (Fitzroy, Vic)*, no. 149 (2017).
22 Turner, *The Green Marble: Earth System Science and Global Sustainability*, 14.
23 Peter Haff, "Humans and Technology in the Anthropocene: Six Rules," *The Anthropocene Review* 1, no. 2 (2014), https://doi.org/10.1177/2053019614530575; V. I. Vernadskiĭ and Mark A. McMenamin, *The Biosphere*, A Peter N. Nevraumont Book (New York: Copernicus, 1998).
24 John Tresch, "Cosmogram," in *Cosmogram*, ed. Melik O'Hanian and Jean-Christophe Royoux (New York: Sternberg, 2005), 74 [emphasis added]
25 Ibid.
26 David Herman, "Cognitive Narratology," in *Handbook of Narratology*, ed. Jan Christoph et al. (Berlin, DE: De Gruyter, 2009).
27 Erin James, *Narrative in the Anthropocene* (Chicago, IL: University of Chicago Press, 2022), 188. See also Erin James, *The Storyworld Accord* (Lincoln, NE: University of Nebraska Press, 2015).
28 David Herman, "Storyworld/Umwelt: Nonhuman Experiences in Graphic Narratives," *SubStance* 40, no. 1 (2011).
29 Mrill Ingram, Helen Ingram and Raul Lejano, "Environmental Action in the Anthropocene: The Power of Narrative-Networks," *Journal of Environmental Policy & Planning* 21, no. 5 (2019).
30 Remember that by "articulate" we mean both to express and to connect (without reducing).

Bibliography

"Gapu-Monuk Saltwater." 2018, accessed August 11, 2023, www.sea.museum/saltwater.

Haff, Peter. "Humans and Technology in the Anthropocene: Six Rules." *The Anthropocene Review* 1, no. 2 (2014): 126–36, https://doi.org/10.1177/2053019614530575.

Haraway, Donna. *Staying With the Trouble: Making Kin in the Chthulucene*. Durham, NC: Duke University Press, 2016.

———. "Carrier Bags for Critical Zones." In *Critical Zones: The Science and Politics of Landing on Earth*, edited by Bruno Latour and Peter Weibel, 440–45. Cambridge, MA: MIT Press, 2020.

Herman, David. "Cognitive Narratology." In *Handbook of Narratology*, edited by Jan Christoph et al., 30–43. Berlin, DE: De Gruyter, 2009.

———. "Storyworld/Umwelt: Nonhuman Experiences in Graphic Narratives." *SubStance* 40, no. 1 (2011): 156–81.https://pakks.org.ec/que-hacemos/.

Ingold, Tim. "Globes and Spheres: The Topology of Environmentalism." In *Environmentalism: The View from Anthropology*, edited by Kay Milton, 31–42. London: Routledge, 1993.

Ingram, Mrill, Helen Ingram, and Raul Lejano. "Environmental Action in the Anthropocene: The Power of Narrative-Networks." *Journal of Environmental Policy & Planning* 21, no. 5 (2019): 492–503.

James, Erin. *The Storyworld Accord*. Lincoln, NE: University of Nebraska Press, 2015.

———. *Narrative in the Anthropocene*. Chicago, IL: University of Chicago Press, 2022.

Kimmerer, Robin Wall. *Braiding Sweetgrass: Indigenous Wisdom, Scientific Knowledge and the Teachings of Plants*. Minneapolis, MN: Milkweed Editions, 2013.

Lorde, Audre. "The Master's Tools Will Never Dismantle the Master's House." In *Sister Outsider: Essays and Speeches*, 110–14. Berkeley, CA: Crossing Press, 1984, 2007.
Morton, Steve, Mandy Martin, Kim Mahood, and John Carty, eds. *Desert Lake: Art, Science and Stories from Paruku*. Collingwood: CSIRO, 2013.
"O Ciclo Annual No Río Tiquié." 2019, accessed July 25, 2023, www.mikkogaestel.com/stage/o-ciclo-anual-no-rio-tiquie/.
Peluso, Nancy Lee. "Whose Woods Are These? Counter-Mapping Forest Territories in Kalimantan, Indonesia." *Antipode* 27, no. 4 (1995): 383–406.
Reynolds, Nedra. *Geographies of Writing: Inhabiting Places and Encountering Difference*. Carbondale, IL: University of Southern Illinois Press, 2004.
Rodriguez, Mariel. "Flowers of Evil/Blümen Des Übels." 2017, https://marielrodriguez.hotglue.me/?flowers.
Rodriguez, Sandy, and Mariana Castillo Deball. "Mixpantli: Contemporary Echoes." December 12, 2021–June 12, 2022, accessed August 11, 2023, www.lacma.org/art/exhibition/mixpantli-contemporary-echoes.
Ruby, Felicity. "Minding the Gap." *Arena Magazine (Fitzroy, Vic)*, no. 149 (2017): 8–11.
"Safecast Map." 2011, accessed August 11, 2023, https://map.safecast.org.
Schneider, Birgit. "Entangled Trees and Arboreal Networks of Sensitive Environments." *ZMK Zeitschrift für Medien-und Kulturforschung* 9, no. 1 (2018): 107–26.
"Terra Forma, Mapping Ruined Soils." 2020, accessed July 26, 2023, https://feralatlas.supdigital.org/?cd=true&rr=true&cdex=true&text=terra-forma-mapping-ruined-soils&ttype=essay.
Tresch, John. "Cosmogram." In *Cosmogram*, edited by Melik O'Hanian and Jean-Christophe Royoux, 67–76. New York: Sternberg, 2005.
Tsing, Anna L. *The Mushroom at the End of the World: On the Possibility of Life in Capitalist Ruins*. Princeton, NJ: Princeton University Press, 2015.
Tsing, Anna L., Jennifer Deger, Alder Keleman Saxena, and Feifei Zhou. *Feral Atlas: The More-Than-Human Anthropocene*. Stanford: Stanford University Press, 2020.
Turner, D. *The Green Marble: Earth System Science and Global Sustainability*. Columbia University Press, 2018, https://books.google.com/books?id=d0pBDwAAQBAJ.
Vernadskiĭ, V. I., and Mark A. McMenamin. *The Biosphere*. A Peter N. Nevraumont Book. New York: Copernicus, 1998.
"The Whole of Acre in a Single Tree: Brazilian History through the Work of the Self-Taught Artist Hélio Melo, a Rubber Tapper from the Amazonas." 2008, accessed August 11, 2023, https://universes.art/en/magazine/articles/2008/helio-melo.

5 Case Study

Global Forest Watch

Global Forest Watch: A Brief History

In 1991, in its second 'WRI Guide to the Environment,' the World Resources Institute (WRI) wrote that their aim was to slow deforestation and save biological diversity. They also expressed concern for people connected to forests, especially when they wrote '[w]hen forests die, so do traditions and livelihoods.'[1] To this day, the institution seeks to 'encourage government [*sic*] to keep forests alive and well and to provide the leadership and funds needed to help conserve tropical forests while making economic development more sustainable and equitable in tropical countries.'[2] Alyssa Barrett from WRI told us in an interview: 'We are a non-profit organization, so we have a theory of change; it's a lot about how we prioritize things, this is dictated by our theory of change so, for example, we decided that our North Star, if you will, is to reduce or stop deforestation, with the emphasis on the tropics.'

Global Forest Watch (GFW) is an online mapping platform that collects an inventory of global forest loss (and in some cases gain) over many years. It offers open data on the status of forest landscapes on a global level with an emphasis on tropical forests. GFW was founded in 1997 when it began as WRI's Forest Frontiers Initiative. WRI has been devoted to environmental concerns with a special interest in tropical forest protection since its foundation as a nonprofit organization in the United States in 1982. The institutional history stands for the two-fold meaning of care as aid and as a power structure, which is a typical relationship between industrialized countries in the temperate zones and 'developing countries,' which often were former colonies, in the tropical belt. This chapter will focus on how this relationship is reflected in the platform.

The partners of the GFW initiative started to publish static forest atlases of countries such as Cameroon, the Republic of the Congo, and Gabon in 1997 because these were places of considerable unregulated deforestation. The static maps became interactive online maps when they were created for Cameroon in collaboration with the Ministry of Environment and Forests of Cameroon in 2004. The technical basis back then was ArcGIS, a family of client, server, and

DOI: 10.4324/9781003376774-5

online geographic information system (GIS) software developed and maintained by the company Esri. These online forest maps became the role model for GFW. In 2006, GFW produced an early global map of remaining intact forests together with Greenpeace (www.intactforests.org/world.webmap.html). Today the digital service is offered as a 'watchdog' to support 'stakeholders in the world's forests—concerned citizens, government leaders, buyers and suppliers of sustainable forest products—who seek to better manage forests and improve local livelihoods' (www.wri.org).

The atlas went interactive and dynamic on a global level when the Web 2.0 tool was launched in 2014 on the platform of Google Maps using the satellite imagery and geospatial datasets made available through the 'Google Earth Engine.' The data included in the online maps of GFW date back to the year 2000, when GPS, which is owned and operated by the United States government as a national resource, was opened up to civilian use.

The long list of past and current funding partners and collaborators of the GFW, including the United Nations Environmental Programme, civil society organizations, international financial organizations, companies, and nongovernmental organizations (NGOs), demonstrate global interest in the tool. The website www.globalforestwatch.org was designed by Vizzuality, an international company specialized in big data-driven environmental monitoring platforms for international organizations following a similar scheme such as Human Rights Watch, Climate Watch, and Global Fishing Watch.

Desktop Analysis of GFW

Our case study of GFW had two phases: a desktop analysis, in which we worked with the platform and mobile app ourselves, applying our research questions from Chapter One and the theory we reviewed in Chapters Two and Three, and an interview component, in which we worked with power users of the platform to see how they used it for their ends and how they dealt with the politics of zoom. The remainder of this chapter details our desktop analysis; the interview component is detailed in Chapter Six.

In the GFW platform, users can apply several different sources of high-resolution satellite imagery and optical sensor data to map layers that differ in spatial and time resolution: (1) annual tree cover change (based on Landsat 5/7/8 satellite data), (2) near-real-time forest disturbance alerts (Landsat, Sentinel 1, and Sentinel 2), and (3) satellite imagery (Landsat 8, Sentinel 2, Google, and Planet).[3] The Planet dataset from Norway's International Climate and Forest Initiative (NICFI) Satellite Data Program, which was set up to cover all the main forest regions of the world with a focus on the tropical belt (www.nicfi.no), is delivered at different time intervals from biannual (2015–20) to monthly (since September 2020). The global layer is refreshed every 10 (Sentinel-2) to 16 (Landsat-8) days. While there might be cloud cover in

Sentinel-2 and Landsat images, for Google satellite imagery cloudy images are omitted to construct a clear view on the planet, but these are not refreshed daily as they are for Sentinel-2 and Landsat. Combined these data sources enable users to visualize global tree cover gain and loss.

GFW integrates different alert systems into the map layers, which allows a set of different perspectives to detect and analyze changes. The GLAD and GLAD-S2 alerts (Global Land Analysis and Discovery), available since 2017, highlight deforestation happening on the basis of days by evaluating Landsat and Sentinel data with the resolution of 30 and 10 meters. The RADD alerts (Radar for Detecting Deforestation) is in function for the humid tropics. Radar optics are able to penetrate cloud cover and to show changes which otherwise would stay hidden on satellite imagery. The Integrated Deforestation Alerts are also applicable for the tropical regions. They are able to detect changes in primary forests, in plantations, and in younger forests.[4] Fire Alerts are an extra layer in GFW. The luminous clusters of red shapes pull their data from two NASA fire sensors, which detect fires at a resolution of 1 kilometer and 375 meters. The sensors indicate temperature anomalies at a given scale, not fire per se (e.g., extremely hot asphalt may also be sensed as fire). The false alarm rate is around 7%. This map layer is particularly difficult to understand from a global perspective without being on the ground or knowing exactly what the vegetation is like on the ground. 'A first glance at the Global Forest Watch (GFW) Fires map shows an aggressive splash of fire alerts across the globe, giving the impression that half the world is on fire,' Sarah Ruiz, Liz Goldman, and Thailynn Munroe wrote for the GFW blog in 2019, when they understood that the world public, sensitized by the fires in the Amazon, might be stunned at more than 100,000 fire alarms presented on their map in a single region.[5]

What Counts as a Tree in GFW?

Having looked into the early history of forest mapping in the second chapter, we learned how forest mapping methods, developed mainly in Europe starting in the late 18th century, coined persistent ideas of trees and forests as monoculture market resources. In general, a single tree is marked on forest maps only if it is rated as a historic landmark. The questions here are how many trees count for a forest in GFW, what size trees have to be mapped, and when a group of trees is large enough to become part of the online inventory from the sky.

To ground-proof the question of what counts as a tree, we took the app known as Forest Watcher to a natural, protected beech forest called Grumsin in Brandenburg, northern Germany. Prior to visiting, we observed the forest from the sky and on our screens using GFW. The different satellite images in GFW present clearly the different types of forests in the region: the regular structure of pine tree plantations planted in rows stood out against the irregular and spongelike texture of the native beech forest. Individual tree species are not discerned by GFW, but it was our local knowledge that enabled us to allocate tree types. Although the area was designated as UNESCO natural World Heritage in 2011, the area is classified in GFW in the following way: 'The region's habitat is comprised of Baltic mixed forests. This region has no Intact Forest. The area has a predominantly warm and temperate climate with high humidity and warm summers. It is part of the Temperate Broadleaf and Mixed Forests biome.' The beech forest has been protected since 1990.

Baltic mixed forests are the natural habitat for Germany, Denmark, Sweden, and Poland. They were defined as ecoregion by the World Wide Fund For Nature, where lowland to submontane beech and mixed beech forests are typical. Experiencing the Grumsin beech forest in situ with the app in our hands made it clear to what extent the experience of being in a forest differs from the bird's eye view in GFW. It was hard to orientate and impossible to locate a specific tree on the app. The Forest Watcher app highlighted some areas with pink pixel structures indicating tree loss, while others indicated the loss of trees through forest fire with brown pixels (https://gfw.global/3YhrxQ4). This allowed us to locate a group of trees logged in the early 2000s, perhaps to expand the established firebreak. And we were able to locate the replanting of beeches in a 4-hectare section where fire and drought had damaged the forest in 2003. In other words, we came to understand just how much experience is needed to connect the understory meaningfully to the overstory.

In comparison to traditional forest maps, it becomes obvious that digital tools mapping trees from the sky make very different judgments as to what counts as a tree or a forest. Foresters counted and measured the size of trees from the ground, the understory, by focusing on single trunks and their diameter. Satellite image systems estimate trees from above by looking at the canopy structure with different cameras using the full spectrum of light and radiation. Consistently, the canopy or overstory view doesn't allow judgments about single trees or tree height. Although historical practices to chart forests and methods in times of satellite imagery, GPS, and online platforms differ drastically, all forest maps tend to blend single trees into the idea of a uniform forest because mixed forests consisting of many tree types are hard to map.

In general, in GFW it's the resolution, texture, and color of the canopy that allow the 'gaze' of the optical satellite sensor systems to classify trees, as the highest resolution available on the analyzable data layers is 10 meters (Planet imagery, which is also available through the platform but is not analyzable, has a sub-5-meter resolution). Instead of individual trees, the satellites detect canopy reflections and changes in the patterns and reflections of the surface. Those changes are quantified as 'tree cover loss' and 'tree cover gain.' So although you might be able to see with your naked eye an individual tree in the Google satellite images, which have the very highest resolution, this tree will not be counted as a gain or loss in the analyzable layers of GFW.

When you start to draw a polygon around an area in the platform, you receive textual information about the type of forest (e.g., 'Baltic Forest' or 'Central Zambezian wet miombo woodlands' or 'no Intact Forest'). Tree cover is defined as 'the density of tree canopy coverage of the land surface.' It is color-coded by density: 'For the purpose of this study, "tree cover" was defined as all vegetation taller than 5 meters in height. "Tree cover" is the biophysical presence of trees and may take the form of natural forests or plantations existing over a range of canopy densities.'[6]

So the question is what counts as a forest rather than what counts as a tree or how to draw a boundary around a forest. Certain decisions have been taken to represent trees in different ways. Some of the different types of forest you can find on GFW are the most biodiverse 'primary forests,' 'intact forest landscapes,' which are 'the world's last remaining unfragmented forest landscapes,' 'tree plantations,' and 'mangrove forests.' Tree plantations are specified into oil palm, wood fiber, rubber, fruit, and mixtures. The platform is closer to a biomass forest calculator than to the mythology of forests in cultures. For example, you can count carbon density by choosing 'tree biomass density,' 'soil carbon density,' and 'mangrove biomass density.' The definition of what counts as a forest in GFW is based on probabilites, not taxonomy. In turn, the heuristic vagueness in definition might allow local definitions and understandings of forests to remain valid. Nevertheless, contradictions, conflicts, and even errors in taxonomy can be identified when a plantation is classified as natural forest or when imported tree species from elsewhere are not considered to be forests for local communities. In Colombia, for example, we found areas color-coded as primary forests that showed, in fact, the gridded structure typical of plantations. In our interviews, we learned about the very different conditions on site, which are homogenized to fulfill the needs of a global forest map. Hence, in GFW, forests are differentiated from other types of land not only automatically by image analysis and deep learning techniques but also through ground proofing and human assistance.

To better understand what is actually mapped in GFW, the question 'what counts as a forest?' can be replaced by the question 'what counts as "environment"?' in GFW. This perspective leads to the following observation: the platform uses forests as a synecdoche—a rhetorical figure that substitutes one part of a problem for the complex whole—for global environment and climate. For example, loading the 'Climate' layer in the main map offers the user the ability to view five different metrics for forest carbon gain/loss and one for soil; in this way, 'climate' is presented in the platform almost entirely as a function of forests. GFW does enable the mapping of other climatic and environmental actors—such as waterways, land cover types, and certain protected species of flora and fauna, but all are presented in terms of their relationships to forests, which concords with GFW's mission statement, 'Forest Monitoring Designed for Action.'

As mentioned before, GFW defines forests based on algorithms applied globally to optical landsat data. Its definitions are therefore, as a general rule, insensitive to local debates over what counts as a forest. For example, some countries draw a distinction between forests consisting of native versus non-native species.[7] But there are also oversights in the monitoring that make them vulnerable to miscategorizations of marginal land cover such as dry (savannah) forests, which were recently found by ground-truthing methods to account as much as 30% of global forest cover.[8] To counteract these results of radical standardization or oversights, GFW does employ ground-truthing in some regions, particularly in central Africa, where the project began.[9] But these efforts rely on third-party funding and are thus not systematically incorporated into GFW's representations of forests. Therefore, forest environments, when they become data for the platform, are subject to a high degree of simplification.

We introduced the concept of the Google Gaia in Chapter Two. GFW becomes most evident as a Google Gaia when looking at the following functions. On the level of forest changes, observers can use GFW as a global forest cinema to monitor different 'patients,' such as primary forests shrinking or suffering from excessive logging. You can play time-lapse videos showing the changes in land cover in any region and on any scale. Or you can calculate forest gain and forest loss and relate deforestation to human action in regard to land use and land rights. To give an example with the data for Nicaragua: from 2001 to 2019, the country lost 460 kilohectare of humid primary forest. This number makes up more than 30% of its total tree cover loss during this same period. In total, humid primary forest in Nicaragua shrunk by 77% in this time period. If you play the time-lapse video starting in 2001, when satellite data became available, you can observe the spread of deforestation in Nicaragua like the spread of a red rash affecting almost the entire country. But GFW doesn't stop there. Users can draw connections between deforestations and different types of concessions such as mining, logging, and plantations, but also between deforestations and global capital. In other words, GFW relates the analytical level with the level of potential action. Many of the country profiles presented on the platform provide the addresses of responsible institutions and regional or national legal information. With the help of this information, users can plan potential action by means of politics and law. But doing so requires an articulation of global and local levels of political action.

The Media Ecology of GFW

GFW is a resource-intense media assemblage of different types of satellites, data centers, and computer devices, but also drones, smartphones, and antennas. In order to act with the data, human monitors have to be equipped with flash disks and smartphones, so that they can connect to the platform, work with the data provided, and upload their own data. Forest data acquired from satellites, such as Landsat or Sentinel, are connected to local forest monitors in the field (communities) via smartphones. The smartphone conditions the evidence on-site via alert (satellite detection), GPS, and cameras as proof. GFW is, in fact, a media assemblage.

When we walked through the German beech forest, we didn't have a reliable connection to the Internet. We had downloaded the maps beforehand in the Forest Watcher app. This is a general problem of remote places, which don't have access to mobile data, which differs geographically and is often scaled up in large contiguous forests. A forest monitor in Peru explained the idea of community monitoring methods:

> But these systems rely on technology access to use. Often the places where the use of alerts could have the biggest impact are deep in the rainforest, with no access to the internet or cell service. 'What good does the information do if it's only seen by a bunch of academics and people in glass buildings?' said Tom Bewick, Peru Country Director for the Rainforest Foundation US, who was a principal architect of the study's monitoring program. 'The whole point is putting it into action.'[10]

To address this general problem, the app also functions offline. Not only can users download the needed material beforehand, they can also pre-upload their local data offline when they are in the field.

Using GFW: Zoom

When you start the GFW map, the well-known Google Mercator world map—with Europe at its center and landmasses represented in white against a blue oceanic background—is provided as the basic interface of GFW to obtain up-to-date knowledge about forest losses and forest gains, protection areas, forest fires, commodities related to forest changes, and climate impacts of forests like carbon losses. The basemap is created from satellite images as discussed earlier. GFW links synoptic views from the sky to local ground by using the zoom tool. By clicking on '+' or '–' users can down- or upscale the map. This is the conventionalized basic function of Google Earth since its beginning.

The aforementioned 'map styles' can be chosen to layer the data (e.g., default, grayscale, satellite, planet, and Landsat satellite images). The interactive map offers manifold different thematic maps (e.g., land cover, biodiversity, mining operations, roads, and political boundaries) that can be added layer by layer to the basic map. Using the zoom tool, it is possible to zoom in from a global view to inspect areas with a resolution of 30 × 30 meters.

Scaling takes place not only on a spatial level but also on a temporal level. The maps contain a timescale starting in the year 2001 and running to the present. The timescale tool can be used to observe certain periods or points of time within the stretch but also as a movie: if one presses ‘play,’ the development of de/reforestation data is played in fast motion, and it quickly becomes clear from the proliferation of pink on the map that more forest has been lost than gained over that time, accelerating noticeably from 2010 to present.

In the default layer selection, tree cover is represented by medium-light green pixels projected on the white landbase map (Figure 2.3). Lines of latitude and longitude, waterways, national borders, and—depending on the map layers loaded—the boundaries of protected areas, palm oil plantations, mines, and managed forests bound and transect the bodies of forests. If any of the satellite imagery basemaps is loaded in the background instead, then roads, landforms, and urban developments form traces on and interruptions in the forest bodies.

At the very center of the GFW platform is the logic of loss and growth; these dynamics are depicted with pink and blue pixels using 30-m resolution, respectively, on the default layer selection. It is important to note here that while average forest loss is readily measured by remote sensing as a function of dramatic changes in the reflectivity of tree cover at the Landsat pixel scale, average forest gain—which produces much slower changes in reflectivity and has less density—is harder to detect and can therefore be underrepresented in at least some locations. This bias means that loss and gain cannot simply be calculated by subtracting the figures; the accuracy of the figures also varies by biome.

The Google Gaia cosmogram is visible in many features of GFW. Land cover and land use are at the center of observing forest change. The platform monitors all changes, caused naturally or by humans, but the main focus is on human impacts, which are the reason for the most severe changes in global forest metabolism and ecology in our time. Consequently, the platform’s main aim is to protect forests from human impacts going forward. The platform makes the argument that deforestation is mainly commodity driven. So, we may say that platforms such as GFW are powerful tools to observe forest changes caused by the exploitative logics of the ‘Capitalocene,’[11] even though the platform doesn’t emphasize this approach explicitly.

Storyworld Networking in GFW (Data and Stories)

GFW has designed a number of ways that users can incorporate their own perspectives, even ones that conflict with the remotely sensed data, into the platform. Users can request to upload their datasets in standard formats and make them either publicly or privately visible as a layer. Through their developer tools on the web platform, GFW makes it possible to create custom maps in MyGFW that integrate the global satellite data collected and visualized by GFW developers with local data uploaded by community users; this is also possible in the Forest Watch mobile app. An example of these sorts of project is the 2S2D project in Cameroon, which used teams of community members to confirm remotely sensed deforestation and to log deforestation that the satellites could not pick up due to lag time or insufficient thinning of the canopy.

Poachers move extremely rapidly through areas, logging at night and moving on before the damage they have done can be registered by a satellite flyover and passed to the GFW team. Meanwhile, local residents patrolling the forests near their houses can detect this activity and alert the authorities. But knowing where to look for new damaging activities can be greatly assisted by the satellite views—just as wildlife conservationists are now using drones to quickly detect poaching activity on game reserves in East Africa. The 2S2D project is a good example of an alternative paradigm to Google Gaia: instead of global views generating global action on deforestation, global views guide the deployment of just-in-time (e.g., 'en temps quasi-reel,' 2s2d.org) spheres of vision and action within local polities in Cameroon.[12] This storyworld network afforded by GFW creates a better fit with the temporal and political scale of action around poaching in African forests.

When we began studying GFW in 2017, there was a story layer to which users could upload stories and photos in their own. In April 2019, due to lack of use and problems with server space and person power, GFW announced a change to *User Stories* driven by the *Places to Watch* feature—in which GFW technicians identify via the remote-sensing data 'hotspots' of deforestation and then conduct investigations or ask local users to upload data on these hotspots. Today, the stories can be addressed through the Mongabay website. The new method tasked a conservation network called Mongabay to carry out this reporting. Now, the only stories shown on the map are in English, and most are written by Western journalists working for Mongabay who either visit the hotspots or report remotely on events happening there—like illegal logging in Senegal or wildlife conservation efforts in Vietnam (Figure 5.1).

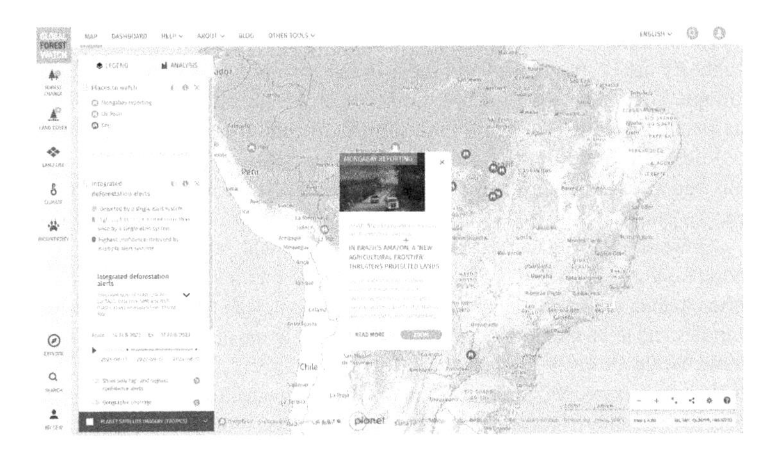

Figure 5.1 Screenshot from GFW with Mongabay stories layer.

Source: Screenshot from http://www.globalforestwatch.org/, August 15th 2023.

This shift is in keeping with the Google Gaia cosmographic narrative, in which transnational organizations are responsible for climate action rather than local actors. However, as part of the *User Stories* feature, GFW does provide accounts of how the platform has been successfully used by local activist groups to catalyze climate action. In '5 Creative Projects Using GFW,' reported on the blog feature of GFW, Sarah Ruiz writes of a project in Oaxaca that mapped several layers of GFW information—including deforestation alerts—over a basemap of the state. The project created this hybrid map in order to empower local *ejidatarios*—individuals and groups entrusted with maintaining the vast system of communal-use land called *ejidos* established by the Spanish during colonization—to make good decisions about land use for their communities: where to plant, where to set guards against illegal logging, where to build or not to build, etc. The project involved building an app that can be used offline on a cell phone and only needs to refresh every 15 days, accommodating the poor internet coverage in remote areas of Oaxaca. The app visualizes GFW data differently than the global basemap does—primarily as heatmaps of deforestation risk. As such, the map is very meaningful to the *ejidatarios* but less so to people who live elsewhere—especially once the map is zoomed into the local level and major city names and the contours of surrounding states disappear. The app's visualizations are therefore dominantly spherical, emphasizing local traditions (the *ejido* system) and matters of concern (illegal logging and land-use management). But the map could not generate these spherical visualizations without global data: as Ruiz explains, 'many *ejidos* do not always have access to the geographic information that could help them make informed decisions about farming and land management.'[13] The Oaxaca project thus constructs a different sort of storyworld network out of the global GFW data—one that expresses and supports the cosmology of the local indigenous population rather than a Google Gaia cosmology. What the Oaxacan visualizations lose in transnational portability they gain in the ability to empower *ejidatarios* to take political action within the scope of their community—a community that has had a longer tenure as 'stewards of the Earth system' in Oaxaca than the transnational organizations that fund GFW.

However, viewers need significant education and resources to get real power out of GFW, that is, to be able to mount multiple layers (which are often hidden in non-obvious menu structures), to upload datasets in standard formats and to be able to read and understand the graph output by the analytic tools. While the GFW mission statement claims that the platform is 'accessed daily by governments, companies, civil society organizations, journalists, and everyday people who care about their local forests,'[14] the tool is used most heavily by academics, NGOs, and governmental agents; this outcome is unsurprising given that local community organizers lacking technological and literacy support may struggle with the complex interface.

The GLAD alerts, the RADD alerts (which are based on radar so that they can look through cloud cover; see Figure 5.2), and Mongabay stories provide another opportunity for situated accounts of climate to challenge global synoptic ones, as seen in GFW maps from regions where recent satellite observations of deforestation overlap the boundaries of federal preserve lands that are supposedly protected from logging. However, the platform definitely privileges the regularly updated remote-sensed datasets in terms of presentation and usability, as remote-sensed layers are much easier to mount from the menu structure than user-uploaded layers.

Finally, users have the ability to customize the synoptic views of forests to meet their local political purposes. For example, they can add layers to the map indicating mining operations and forest loss and then compare them; they can then search over those layers by time and region (country/jurisdiction or user-drawn polygon); they can summarize forest change over their chosen time/space by using the analysis tool; and they can set their own parameters for canopy density to better reflect local definitions of what counts as a forest. However, data analysis is limited to tree gain/loss and cover, and users cannot introduce new variables or search datasets using their own identifiers.

Conclusion of the Desktop Analysis of GFW

As reviewed earlier, users of GFW have several ways they can complement or counter global views of their forests of concern with local views. In

Figure 5.2 Screen capture of GFW with RADD alerts in the Central African Republic based on Radar data by Sentinel 1 within 10-meter pixels. The lines indicate recent logging in a typical grid-like structure along roads.

Source: Screenshot from http://www.globalforestwatch.org/, August 15th 2023.

MapBuilder, federal land managers can upload country-specific datasets and use GFW's tools to calculate total forest gain and loss, and these results can be compared with remote-sensing data. Users can subscribe to GLAD, RADD (integrated) deforestation alerts, and VIIRs fire alerts, allowing community activists to draw on the assistance of remote sensing and crowd-sourcing to stop illegal activities in their areas. And users of the Forest Watcher mobile app can upload local data and photos for viewing in tandem with the global datasets. All of these features combine global and spherical views of forest climate in ways that help local communities take meaningful climate action, as exemplified by the palm oil interventions reported on the WRI site.

However, downscaling technologies continue to privilege globalized, neoliberal actors in GFW: academics with technical training and multinational funding; government employees who use remote sensing to surveil and control traditional forest use by communities in their jurisdictions; transnational mapping/surveillance companies like Google who may harvest user-supplied data for their own purposes; and corporate sponsors such as Cargill and Unilever, who seek to change (at best) and greenwash (at worst) environmentally unfriendly practices associated with the booming palm oil industry. These uses and perhaps abuses of power cannot be administered by GFW. However, the organization can tip the scale away from totalization and toward empowerment by strengthening the storyworld networking capacity of the platform. And in fact, as we saw in our interviews, many local users of GFW are coming up with ingenious solutions to create storyworld networks of their own.

Notes

1. Kenton Miller and Laura Tangley, *Trees of Life: Saving Tropical Forests and Their Biological Wealth* (Boston: Beacon Press, 1991), ix
2. Ibid.
3. "Using Satellite Imagery to Investigate Deforestation on Global Forest Watch," *Webinar*, posted by "Global Forest Watch", July 28, 2021, www.globalforestwatch.org/help/map/webinars/satellite-imagery-investigate-deforestation/.
4. Anika Berger, Teresa Schofield, Amy Pickens, Johannes Reiche and Yaqing Gou, "Looking for the Quickest Signal of Deforestation? Turn to GFW's Integrated Alerts," *Global Forest Watch* (blog), March 9, 2022, www.globalforestwatch.org/blog/data-and-research/integrated-deforestation-alerts/.
5. Sarah Ruiz, Liz Goldman and Thailynn Munroe, "Placing Global Wildfires into Local Context," *Global Forest Watch* (blog), September 6, 2019, www.globalforestwatch.org/blog/fires/placing-global-wildfires-into-local-context/.
6. See the definition on GFW, accessed through Global Forest Watch, following Matthew Hansen et al., "High-Resolution Global Maps of 21st-Century Forest Cover Change," *Science* 342, no. 6160 (2013): 850–53.
7. Paul Robbins, "Fixed Categories in a Portable Landscape: The Causes and Consequences of Land-Cover Categorization," *Environment and Planning A* 33, no. 1 (2001): 161–79. https://doi.org/10.1068/a3379.
8. Betsabé de la Barreda-Bautista, "Tropical Dry Forests in the Global Picture: The Challenge of Remote Sensing-Based Change Detection in Tropical Dry Environments," in

Planet Earth 2011—Global Warming Challenges and Opportunities for Policy and Practice, ed. Elias Carayanni (Tech: Open Access Publisher, 2011), 231–256, DOI: 10.5772/24283.

9 Brittany L. Peterson, "Thematic Analysis/Interpretive Thematic Analysis," *The International Encyclopedia of Communication Research Methods* (2017): 1–9.

10 Global Forest Watch, "Supplied with Tech, Indigenous Forest Monitors Curb Deforestation," *Global Forest Watch* (blog), July 12, 2021, www.globalforestwatch.org/blog/people/indigenous-forest-monitors-reduce-deforestation/.

11 Jason Moore, *Capitalism in the Web of Life: Ecology and the Accumulation of Capital* (London: Verso, 2015).

12 "Système De Suivi De La Dégradation Et La Déforestation De Cameroun," 2023, accessed August 29, 2023, www.2s2d.org.

13 "5 Creative Projects Using Gfw to Protect Forests." 2019, accessed November 4, 2020, https://blog.globalforestwatch.org/people/5-creative-projects-using-gfw-to-protect-forests/.

14 "About," *Global Forest Watch*, 2023, accessed August 23, 2023, www.globalforestwatch.org/about/.

Bibliography

"5 Creative Projects Using GFW to Protect Forests." 2019, accessed November 4, 2020, https://blog.globalforestwatch.org/people/5-creative-projects-using-gfw-to-protect-forests/.

"About." *Global Forest Watch*, 2023, accessed August 23, 2023, www.globalforestwatch.org/about/.

Barreda-Bautista, Betsabé de la. "Tropical Dry Forests in the Global Picture: The Challenge of Remote Sensing-Based Change Detection in Tropical Dry Environments." In *Planet Earth 2011—Global Warming Challenges and Opportunities for Policy and Practice*, edited by Elias Carayanni, 231–56. Tech: Open Access Publisher, 2011, DOI: 10.5772/24283.

Berger, Anika, Teresa Schofield, Amy Pickens, Johannes Reiche, and Yaqing Gou. "Looking for the Quickest Signal of Deforestation? Turn to GFW's Integrated Alerts." *Global Forest Watch* (blog), March 9, 2022, www.globalforestwatch.org/blog/data-and-research/integrated-deforestation-alerts/.

Global Forest Watch. "Supplied with Tech, Indigenous Forest Monitors Curb Deforestation." *Global Forest Watch* (blog), July 12, 2021, www.globalforestwatch.org/blog/people/indigenous-forest-monitors-reduce-deforestation/

Hansen, Matthew et al. "High-Resolution Global Maps of 21st-Century Forest Cover Change." *Science* 342, no. 6160 (2013): 850–53.

Miller, Kenton, and Laura Tangley. *Trees of Life: Saving Tropical Forests and Their Biological Wealth*. Boston: Beacon Press, 1991.

Moore, Jason. *Capitalism in the Web of Life: Ecology and the Accumulation of Capital*. London: Verso, 2015.

Peterson, Brittany L. "Thematic Analysis/Interpretive Thematic Analysis." *The International Encyclopedia of Communication Research Methods* (2017): 1–9.

Robbins, Paul. "Fixed Categories in a Portable Landscape: The Causes and Consequences of Land-Cover Categorization." *Environment and Planning A* 33, no. 1 (2001): 161–79, https://doi.org/10.1068/a3379.

Sarah Ruiz, Liz Goldman, and Thailynn Munroe. "Placing Global Wildfires into Local Context." *Global Forest Watch* (blog), September 6, 2019, www.globalforestwatch.org/blog/fires/placing-global-wildfires-into-local-context/

"Système De Suivi De La Dégradation Et La Déforestation De Cameroun." 2023, accessed August 29, 2023, www.2s2d.org

"Using Satellite Imagery to Investigate Deforestation on Global Forest Watch." *Webinar*. Posted by "Global Forest Watch", July 28, 2021, www.globalforestwatch.org/help/map/webinars/satellite-imagery-investigate-deforestation/.

6 Insights From Developers and Users of GFW

In the previous chapter, we related our desktop critique of GFW, in which we concluded that (a) because of its foundation in the Google Earth Engine and its ties to transnational organizations—both corporate and non-profit—GFW does replicate a Google Gaia cosmology of forests, which by default encourages Global North solutions to Global South problems; and (b) nonetheless, its open-source architecture, its openness to local data sets, its storytelling and blogging features, and the development of the mobile Forest Watcher app, MyGFW, and Map Builder for users interested in data sovereignty—these features create promising conditions for storyworld networking.

We wanted to know if those promises were borne out in users' actual experience. So, we conducted interviews with three developers/administrators of GFW and four power users of the platform who operate at various global sites (all interviews/protocols were conducted using an Internet video-conferencing platform under IRB/Ethics Board approval, protecting the anonymity of participants to the degree they requested). For three of the power users, we additionally asked them to complete think-aloud protocols, during which we watched them perform typical tasks in GFW while providing commentary as they worked; this procedure allowed us to capture gestures and patterns working with the software—important information as one of our main questions in this book is about the political implications of zooming in a global platform.

Our three developer interviews were with the following people at or involved with GFW:

- David Gonzalez, one of the co-founders of Vizzuality, the web design company that contracted with WRI to build the GFW platform;
- Alyssa Barrett, current product strategy lead for the Land and Carbon Lab at WRI, but the original product manager for GFW. In that role, she was involved in most design and development decisions, and she oversaw the initial user testing and the resulting feedback process with the web developers;
- Jessica Webb, strategy lead for People and Forest Protection at WRI. She interfaces with stakeholders to help them identify the best GFW products

DOI: 10.4324/9781003376774-6

for their monitoring and management needs, identifies gaps in forest data and works with the development team to fill those gaps, and addresses human dimensions, equity, and sustainability in the use of GFW tools for forest management.

We aimed to secure a diverse range of GFW user site case studies across the globe. Our case studies were at the following sites, some of which are referred to only in general terms to protect participants doing sensitive work in those regions:

- Indonesia: We studied a nongovernmental organization (NGO) that uses GFW to stop deforestation in Aceh, Indonesia. They were early adopters of the forest disturbance alerts and helped pilot early versions of Forest Watcher. While forest preservation is a federal responsibility in Indonesia, the government often lacks funding, tools, and resources to adequately monitor and manage an intensely forested nation: the country is over half forest, and half of that is old-growth. In addition, Indonesian forests, particularly in the Aceh region, exhibit rare levels of biodiversity and house multiple keystone endangered species, including tigers, elephants, and orangutans. The greatest threats to Indonesian forests are currently illegal plantations, fuel removal, and pollution, due to an ongoing economic crisis following the 2004 tsunami; according to the UN Food and Agriculture Organization (FAO), Indonesia lost roughly 20% of its forest carbon stocks to these stressors between 1990 and 2010.[1] The NGO we studied has helped provide GFW tools and training to two major forestry agencies in Indonesia to help them preserve and recover Indonesian forests. This study site helped us consider the challenges in visualizing complex forests and the problems articulating views of forests among NGO, governmental, industrial, and Indigenous stakeholders.
- Georgia: We worked with Gigia Aleksidze, an analyst with the Georgian Ministry of Environmental Protection and Agriculture. The ministry engaged in an intensive project with GFW starting in 2015 to build a custom platform incorporating GFW's global datasets with Georgian land atlas information. The ministry is in the process of expanding the datasets to include agricultural and mining data as well as incorporating historical Soviet records of forestry and land use. However, a cyberattack in the spring of 2020 put a damper on those projects; when we spoke with Aleksidze in the summer of 2022, the ministry was scheduled to relaunch the rebuilt platform in the fall (at this writing, it is back up and running). Georgia is almost 45% forested, largely with deciduous montaine forests, and has an interesting historical situation in that during its membership in the Soviet Union, it was forced to import forest products from Russia, leaving its own forests relatively untapped during that period. With independence, Georgia gained control of its forests, and along with the benefits of that repatriation

came the stresses of illegal logging for construction, water distribution, overgrazing, and resulting erosion. Georgia's steep topography is particularly vulnerable to landslides. This was a study site for us in that it treated a northern temperate forest with good governmental oversight that still experiences significant challenges negotiating forest uses among public, private, and civil sectors.

- Cameroon: We worked with Phanuella Djanteng, Cameroon project manager for SAILD (Service d'Appui aux Initiatives Locales de Développement) and since April 2023 an employee of WRI. SAILD was one of GFW's first development partners and provided test sites for integrating local forest information with global information in the platform. Djanteng works with communities and monitors on the ground in and around Deng Deng National Park using the Forest Watcher app; she also reports emergent and ongoing forest-use conflicts to relevant governmental agencies. Estimates of Cameroon's forest coverage vary from 40% to 60%. Nearly half of that is rainforest. Stressors include illegal logging and poaching of other forest products (herbs, animals, etc.) for trade. Increasingly, climate change-related wildfires are also taking a toll. Nevertheless, Cameroon's growth conditions remain so favorable for trees that reforestation efforts have been successful, reducing the country's total tree cover loss to 2% since 2000. Djanteng reports that greater concerns include the welfare of human communities around protected forest areas, given economic hardships with the devaluation of the Central African franc, and species diversity within forests. This was a good study site for us in that it spoke directly to the issue of people being an often-invisible part of forests.
- Peru: We worked with members of an NGO that uses GFW to monitor deforestation in and around Indigenous lands caused by illegal logging, mining, and plantations. This NGO works with Indigenous monitors, primarily via the Forest Watcher app, to ground-truth GFW's integrated deforestation alerts. When possible, the NGO exports waypoints for ground-truthing and sends them as data to monitors' cell phones. But due to poor connectivity in remote regions, the NGO also expends considerable time and money exporting maps from GFW as slideshows that are hand-carried into the forest on laptops or as printouts. Increasingly, the NGO is training Indigenous monitors to use drones to monitor deforestation areas that are dangerous due to illegal plantation and drug trafficking activity. Peru is just over half forested with the vast majority of this area being primarily montaine forest with high biodiversity. Heavy dependence on mining in the country has led to illegal prospecting in protected Indigenous lands and peripheral drug trafficking and bio-piracy in response to economic pressures. This study site highlighted the challenges many South American Indigenous peoples face in trying to maintain their traditional lifeways and territorial rights in forests.

Interview Analysis: Articulating Global and Local Views of Forests in GFW

To analyze our protocol interviews, we chose a common rhetorical methodology: thematic analysis, which works iteratively and inductively to surface dominant themes across open-ended responses from participants.[2] Applying this method, we found that four main themes emerged around the problem of *articulation*—namely, the problem of *how to get scales, media, and communities to work together when they can't simply be reconciled or reduced to each other*. These four points of articulation significantly echoed the themes that emerged from our desktop analysis of GFW: the problem of zoom, the wood wide web (media ecology of GFW), the canopy problem, and maps as boundary objects. We'll discuss each theme in turn, focusing on problems users reported and solutions they engineered.

The Problem of Zoom

Making Local and Global Data Play Well Together in GFW

Our interviews with developers and managers made it clear that this had been a problem since the design stage of GFW. First of all, David Gonzalez recalled a specific technical problem with pixel size:

> I remember a 50–60 message email thread I had with someone at Google about how to represent deforestation globally. This is a kind of a technical thing but it's a good example of how much we thought about this. So if you see the map the global map you see all these all those pink dots and when we first presented it someone said, this is incorrect looks like half of the world is deforested. . . . That can't be. This is wrong. Because when you are at zoom number one, when you see the whole world in your screen, one pixel is more or less 6000 square kilometers, right? So if you were to say that 6000 kilometers are completely deforested, the first picture would be pink. If only 3000 square kilometers are deforested, it would be half pink. If only 500 square kilometers are deforested, you wouldn't see the deforestation at all in that spot. So turns out, if you do it correctly, it looks like there's no deforestation in the world at all. It is correct, from a visualization theory perspective, but it's not telling you anything. And this was a real struggle. It's like okay, we want to be scientific, but I mean we have to make some decisions. I will have to convince this [scientist] that we are going to be forced to exaggerate things a bit when you are zoomed out because we don't think people are going to do thorough analysis at that level. As you zoom in, it becomes more and more accurate.

Gonzalez also recalled there being a general tension with incorporating really accurate local data into the platform due to server capabilities. Alyssa Barrett brought up this tension in her interview as well: 'We're a global-oriented project; we don't have the capacity to go to every single country and gather data, and . . . sometimes the most useful data is at a very local scale . . . . It's not only collecting it once, but you also have to commit to making sure it's up to date . . . . So . . . we prioritize certain geographies.' This issue led the development team at WRI to partner initially with agencies in a smaller number of countries with at-risk tropical rainforests, like Cameroon and Indonesia. But Barrett also emphasized that it was important that GFW provide a global overview to articulate various transnational projects with each other:

> At the end of the day, the focus of Global Forest Watch is to provide access to the best available globally and temporarily consistent data . . . . By design the tree cover loss data is consistent everywhere, so that you can compare apples to apples. And . . . we know that that's not the one and only way of looking at forest . . . . Still, I mean carbon markets are exploding right now, but nobody has good dataset baselines and there's not a lot of trust, so we're trying to help with that [problem] in our own way.

Another problem with articulating global and local datasets that came up in our interviews was the problem of data quality. This was something Gigia Aleksidze had particularly struggled with in his work with the Georgian MEPA:

> For instance . . . we got some of the datasets about mining and about deposit sites from . . . a separate Ministry. But when we integrated this data, we found out that there were so many errors. . . . We needed to clean up this data and make some changes. And even now, we haven't found out all the errors which [still] exist in that dataset. That's why I'm saying that it also depends what quality data you have integrated.

Still another issue that was raised by our participants was the problem of articulating temporal scales. As we saw in Chapter Three, Zachary Horton argues that zooming is both a spatial and a temporal process.[3] We observed our participants working to articulate their understanding of deforestation between these two processes—quite literally in our Cameroonian case study, as Djanteng told us she was looking back and forth between the map and the timeline as she played the deforestation data forward across her three-month period of interest, to see which areas were most affected during that time. Jessica Webb further noted in her interview that users had initially struggled with the timeliness of the GFW datasets:

> Global Forest Watch launched with the flagship data set from the University of Maryland, which is the annual tree cover loss. That is very good at being able to detect at a relatively granular level, for example to analyze how deforestation trends and how forests are changing over time. But

> because they're annual, and we get the data from University of Maryland for the previous year like in March or April, sometimes there can be a lag of over a year between when the deforestation or tree cover loss actually happened [and when it appears in the dataset]. Which isn't entirely enough for people who are focusing on the local level.

These people are the cultural guardians and law enforcement agents who need to know about illegal deforestation in time to limit it. Webb noted that this problem with temporal articulation led GFW to work with researchers to develop near-real-time forest disturbance datasets, based on Landsat and Sentinel data, that are updated on a daily basis on GFW.

Another problem Webb noted was with the overwhelming amount of data from global and local sources—not only from the developer's standpoint of having to serve and integrate the data but also from the user standpoint and not knowing how to visually parse all the overlapping layers of alerts from different remote-sensing datasets. In our protocols, we observed both our Cameroonian and Indonesian participants deselecting all layers except deforestation and VIIRS (fire) alerts in order to simplify their views of their areas of interest, which confirms GFW's decision to commission those alert layers to serve local users. Webb further remarked that 'just very recently, the last couple months based on user interviews, we decided to streamline the three [GLAD & RADD] systems and present them as one layer on Global Forest Watch which we're calling integrated alerts.' All four of our case study sites mentioned using the integrated alerts so that development is clearly beginning to have an impact. As a matter of fact, in Figure 6.1 we see Djanteng using the integrated alerts while zooming into her forest of concern (FOC) in a national park; the final screen capture in the series shows her viewing the certainty analysis on the alerts displayed within her selected polygon.

A final problem with integrating global and local data came from our Peruvian site—particularly around the issue of being able to interpolate local maps of Indigenous protected lands with GFW alerts. They noted that 'GFW is a platform that lets you export or download data, but it's very difficult to incorporate data. If you want to do a fuller analysis with your own layers, it's very hard to do that.' Our Indonesian participant also mentioned that for certain kinds of analysis he had to move into another GIS app: for instance, he couldn't calculate tree cover loss from the integrated alerts for his FOCs in hectares, the units that Indonesian governmental agencies use.

If these were the problems with articulating global and local views, developers and users had also come up with solutions. First, we noticed that users navigated the problem of articulating discontinuous images of their FOCs at different scales by choosing a stable reference point against which to compare the changing views. In the think-aloud protocols, at least two of our participants positioned their mouse pointers over a landmark of interest and then used the track/roller on their mouse to zoom in or out around that

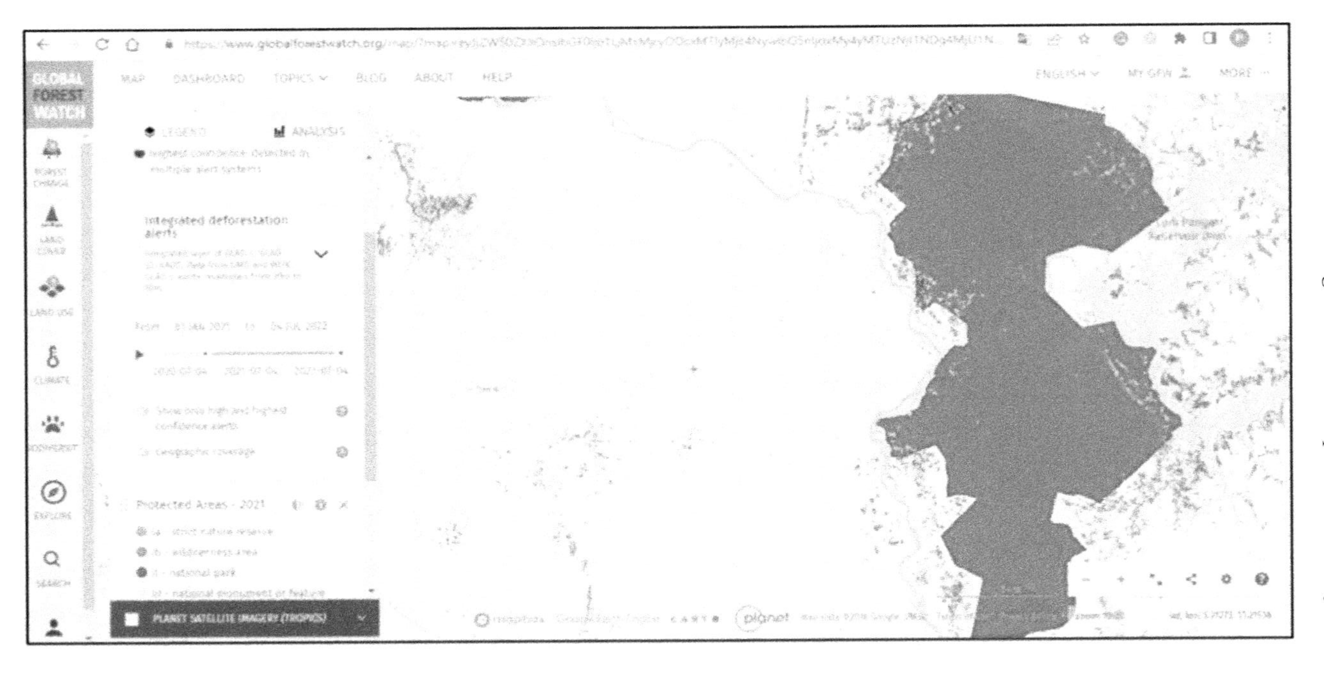

Figure 6.1 Sequential screen captures of zooming activity in Cameroon around a national park (dark area) and using integrated alerts (grey dots).

Source: Figure adapted from Screenshot from Zoom session using http://www.globalforestwatch.org/, July 5, 2022.

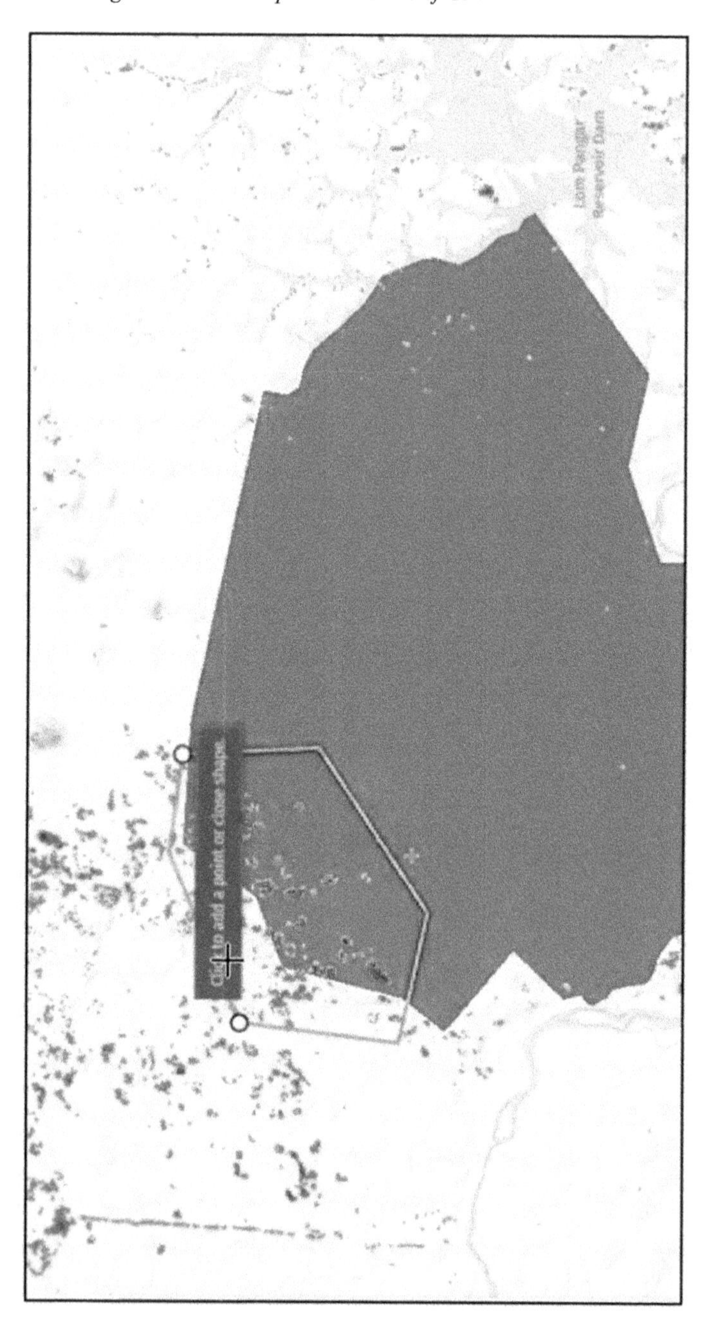

Figure 6.1 (Continued)

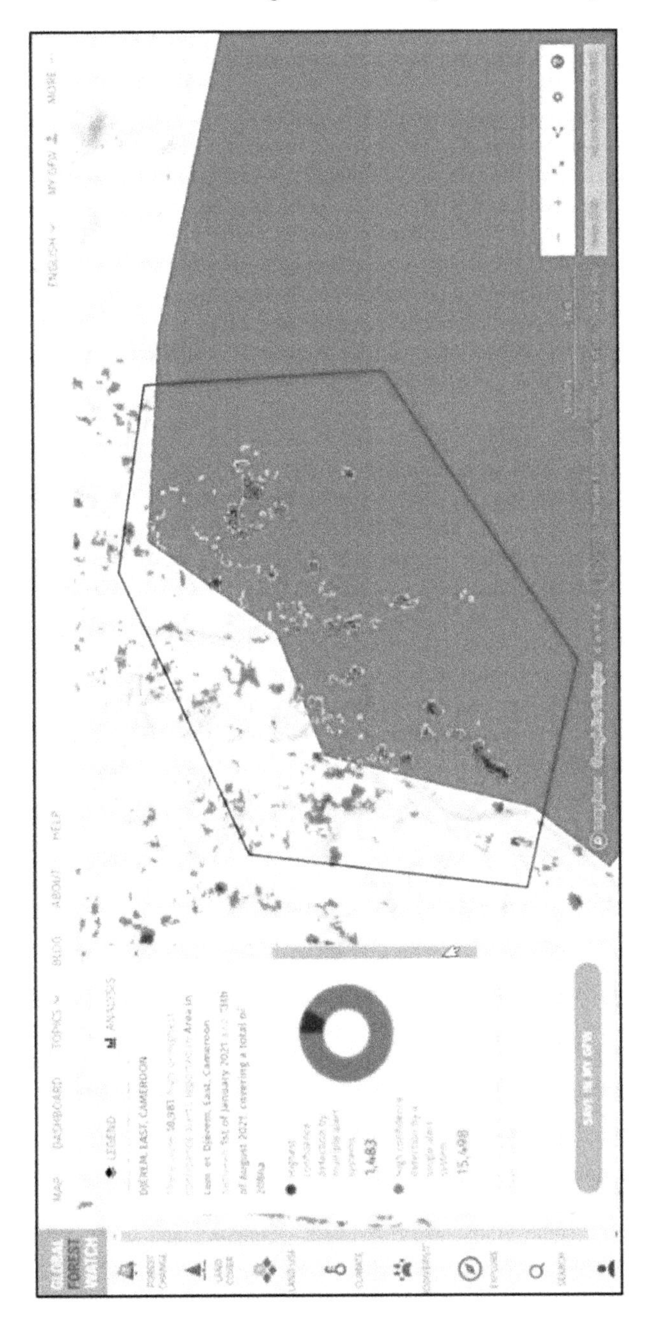

Figure 6.1 (Continued)

point. Another solution users developed was ground-truthing the pink deforestation pixels in GFW with human or drone monitors—more on that in the next section.

Other solutions to articulating global and local views of forests appeared to derive from recent data sovereignty movements, in which local communities—particularly Indigenous communities—are resisting uploading their data into a globally accessible cloud, seeing that move as risky for community safety and a continuation of colonial extraction of Indigenous resources. Gonzalez specifically remarked on this move toward localization as being a possible solution to the problem of the incommensurability of global and local views: 'Maybe we don't need global datasets for everything.' Barrett noted that GFW had built out Map Builder as a solution for users who wanted to run GFW on their own servers with their own datasets. Webb commented:

> Particularly with the community monitoring, it also been very disappointing in the past, where they had worked with NGOs and doing monitoring exercises and other things, and then they gave that data to the NGO and then just never heard anything again, and it wasn't helpful to them, and so the ownership was a really important piece. So, we decided when we built the Forest Watcher application, as well as the My GFW login, which allows users to upload their own shape files or save areas that they've drawn that that was accessible to them so they could, you know, re-analyze that data, show that data, they can interact with it, they can share it [within] their organization privately and securely, without it being available to others unless they choose to share it.

At the same time, Webb noted that some Indigenous communities were galvanized by seeing their traditional forests, cultural sites, and villages from space in GFW:

> I've also heard that, particularly for communities who often feel . . . forgotten or like disregarded or not important or people don't care about the issues that they're facing, [they] say, 'Oh my gosh, you can see us' and . . . you can see the illegal deforestation that this company or this person or whatever is causing on the map. You know . . . it feels very empowering to them to know that this information exists. It's backing up what they're saying and validating that. And they're on the map.

These are some of the ways in which developers and users of GFW have articulated seemingly incommensurable scales to accomplish their environmental justice (EJ) goals. Other solutions to the problem emerged under the theme of ground-truthing global views of forests.

Validation of Global Alerts by Local Ground-Truthing

In the last section, we saw that a significant problem for both developers and users of GFW was the fact that despite the power of the zoom tool, users have additional forest visualization needs that exceed the temporal and spatial scales afforded by the platform.

One aspect of the problem is the certainty of the information. Our Peruvian site reported that the Planet imagery was often good enough to confirm if a forest change alert was due to an illegal plantation. But our Indonesian study site gave lower odds during his think-aloud protocol:

> I can reduce the opacity and change the layer and then we can also reduce the opacity to see the Planet[data]. So we can zoom in and check the alert as a second step. . . . So on the alerts its 50%, and then with the PLANET or other satellite imagery like Sentinel 2, we can say that we have 80% confidence that it's really happening in the field, something like deforestation or landslide.

As hinted at in our Indonesian participant's response above, another major user need in this respect is determining the *cause* of the deforestation; sometimes, in the case of wildfires, that cause is at least partially visible in GFW, but more often it is not. Webb explained, 'Because of course the satellites and algorithms don't detect [human-caused] deforestation, they just detect some sort of change, so sometimes if there's like flooding, you know, a wind-blow event or a landslide or something like that, it's not human caused.' The obvious and necessary next step for all of our sites was ground-truthing—sending human informants to the location of an alert, or as close as was safe for them—to confirm the nature and cause of the forest canopy change visualized in GFW. Webb remembered going out with a field team in the Congo:

> I mean there's been a few times . . . seeing alerts in the middle of pristine forest, protected area, and it's just like there is no way that's actually something. But if we're in the area anyways we'll go, or if patrol groups are going there any ways to go check it out. And I remember one time I was in the Republic of Congo, we spent, I don't know, a couple hours hiking through this forest and then, sure enough, even being you know a few meters away, you can't see [the clearly because] it's so dense. And we walk into this clearing that somebody had cut down to plant something in the middle of nowhere, and . . . the forest monitors had no idea. So,

> I think it's moments like that where they're like, 'Oh wow, okay, this is really useful.'

Our Indonesian participant described a similar process of delivering alert coordinates to monitors for ground-truthing:

> We have 13 people in the field in the surrounding areas. They will check our information and then they will collect the data, including the photos and the GPS coordinates and also the tally sheet for the information to include in the database. So after they're back from the field, the database person will do the entry of the data and then we will visualize the data from the field along with the deforestation . . . layer in the Google Maps platform; we will connect all the information of GPS coordinates with the photos and also the database for its location. So we come [back with] 100% in-the-field confidence that something happened, and that's some power.

Similarly, ground-truthing generates a richer context that can support more confident inferences about the causes of deforestation. Djanteng told a particularly interesting story about how SAILD went into the field to check an alert but didn't initially locate the problem. It was only after speaking to a local farmer that they located the swath of dead trees they had seen in GFW. The farmer himself had been perplexed by the die-off, but Djanteng's team was able to deduce from the evidence that the trees had been drowned by overflow flooding from a new dam project; they took pictures and sent a report to the governmental agency in charge of the dam. The NGO and the local community needed to work together to solve a problem that neither of them would have been able to solve alone. In another case, SAILD discovered on several field missions that patchwork deforestation in a protected area was actually small subsistence farms cleared by local landholders who had been dispossessed by the dam project and had nowhere else to turn. This was another occasion in which ground-truthing uncovered causes of deforestation that SAILD could communicate to public authorities to promote EJ action—in this case, support for the dispossessed farmers.

As is apparent from these anecdotes, ground-truthing not only solves the technical problem of articulating global and local views of forests with more certainty, but it also strengthens political connections among stakeholders working on forest preservation. Our Indonesian, Cameroonian, and Peruvian informants all stressed that they never carry out monitoring work without the cooperation of local Indigenous communities. Djanteng noted, 'When we go to an area the first time we get in touch with the local authorities there. We present ourselves. We present the purpose of our visits. We get in contact with the traditional leaders.' In other words, they articulate themselves to the local networks that protect the forests.

The Wood Wide Web: Media Ecology of GFW

This theme emerged from developers and users talking about the challenges of articulating various visualizations in GFW—both across the various apps within the system and between GFW and other media and display systems. Again, participants discussed both problems and ingenious workarounds—all of which demonstrate the multiple articulation joints of any media ecology, no matter how seamless it might appear from the outside.

Disconnects Between Intended and Actual Users

Developers in particular faced the problem of articulating their intentions for GFW's use with the users who could actually tap into its media ecology. GFW has a mission statement that includes a 'theory of change,' according to Barrett: 'our North Star, if you will, is to stop deforestation, with an emphasis on the tropics.' Notwithstanding, when GFW does user surveys, they have found that their user group is dominantly Global Northern white males. This is largely because GFW is at its core an interactive Internet-based application built on the Google Earth Engine: that architecture sets constraints immediately in terms of who will be able to access and exploit the media ecology—namely, people with good Internet connections, fast computers, and a good technical education. Gonzalez and Barrett both acknowledged this conundrum, with Gonzalez emphasizing that 'we don't want Global Forest Watch to be a tool used by white male analysts in DC' but admitting that it was a constant struggle pushing through the privilege required to access GFW's media ecology. Barrett noted that due to these constraints, 'we often rely on our international offices sort of as a proxy for end users' when they test apps and features. But, she pointed out that this problem was why GFW had developed the Forest Watcher app in the first place—specifically to make the platform more usable by forest monitors in the Global South.

Aside from Internet access issues, which we'll turn to in a moment, technical education remains a significant barrier for non-Western-educated forest guardians who want to use the mobile app. Several participants on the GFW development and management team acknowledged this problem, including Gonzalez:

> Some of our developers, designers, etc. went on a trip to Peru with some people who used this [Forest Watcher] type of app and it was actually very enlightening in many ways, you know. Because . . . we have . . . preconceived ideas of how . . . . you should make a mobile app. Like okay, we all know, the hamburger icon means menu . . . and [suddenly] we're talking to people who may have never used an iPhone app or an iPhone . . . before.

The solutions to this problem of articulating GFW's intentions with its user base across its media ecology have been two-fold: first, training and, second, shifting emphasis in user-base development. First, GFW has conducted field training with all its test sites and continues to offer trainings whenever new apps or products roll out. NGOs and agencies who use GFW products also conduct their own trainings with Indigenous monitors. Webb discussed some surprises she encountered while conducting these trainings: 'But even for the Indigenous communities that are fairly remote . . . the younger folks just pick it up so quickly, and even in the years I've been doing this [since 2015], like . . . in Latin America and the Amazon just the number of [Indigenous] people who have cell phones now . . . it's quite prevalent, so I think it's just more familiar to people as they have access to these kinds of apps.' And these Indigenous folks often pick up the technology well enough to teach others: for example, our Peruvian participants reported training local schoolteachers in Indigenous communities in Forest Watcher and then having them pass that information on successfully to monitoring teams; also, they've begun teaching Indigenous women to fly drones as a way of enhancing their access to technology as well as keeping monitors safe from retributory attacks by poachers and drug traffickers. So, training as a solution to mission articulation problems across GFW's media ecology has resulted not only in better access of remote Indigenous communities to those ecologies but also in better appreciation by GFW developers of the skills and resilience of Indigenous communities.

At the same time, however, GFW is responding to the somewhat intransigent problem of the 'white male analyst in DC' by leaning into it: they've developed GFW Pro and are actively marketing it to private-sector companies who need more power and privacy in their analyses. Barrett was most cogent on the reasons for this shift as she discussed the need to respond to 'exploding' carbon markets with reliable baselines for calculating futures. She also argued, 'we need to improve transparency in supply chains, reduce the footprint of agricultural products and forest products.' Those supply chains are made up of 'the company and the commodity buyers and traders, too. So the companies are sort of the top and then there's a whole web of actors beneath them. I think other groups, too, like rating companies and the Greenpeaces of the world.' Results of an internal 2021 GFW user survey, to be discussed in more detail below, indicate that Barrett's intuitions are correct; the survey reports a 'significant' base of users interested in forests primarily as carbon capital.[4]

Within the GFW universe itself, there are non-trivial problems moving visualizations across the different apps and the devices they run on. As mentioned earlier, Aleksidze reported that a cyberattack had rendered the Georgia Forest Atlas (a custom MapBuilder site incorporating GFW and local datasets) inoperable for his ministry; when we interviewed him, his team was having to reconstruct parts of the system and reload licenses in order to operate it again. But even when the ecology is functionally normal, there are barriers to the fluid circulation of visualizations. One issue is the different screen sizes between a desktop and a mobile device. Gonzalez pointed out, 'When you're standing there under the canopy in the dark with your smartphone in your hand, you don't need a little pointer arrow like you have in Google Maps, you need a big, huge arrow blinking at you and telling you which way to walk in the forest.' He thus alluded to the problem of reconciling layers of information in GFW, particularly location data (i.e., topographic features on the map) and deforestation data (pink pixels). We observed our Indonesian participant negotiating this mismatch during his protocol by turning the opacity of the deforestation layer up and down as he worked to get more accurate estimates of location.

Gonzalez further pointed out that depending on how much local data users want to look at, there can be cross-platform issues:

> At times we've had like four platforms working at the same time in GFW: RDS [Relational Database Service], CARTO, Earth engine and MapBooks at the same time. And still some of the layers may come, like, for example . . . we have the Brazil indigenous lands, but the government said they are not going to send us the file because they have a policy that they serve it from their RDS servers. So, we have to connect to the RDS servers.

Both our Peruvian and Indonesian participants reported problems with exporting maps from GFW to Forest Watcher, or maps from Forest Watcher to print or to display in other applications.

The solutions to these problems were generally work-arounds that exploited peripheral parts of the media ecology to move visualizations around the system: the two most common appeared to be dropping pins on alert areas and exporting those coordinates as GPS data across lo-fi data connections to cell phones, where they could be interpreted by native mapping apps and navigated to by monitors, and printing out GFW maps, sometimes on multiple pieces of letter-sized papers that were then taped together and carried to field sites where they could be posted on a wall or laid out on a table and viewed by monitors. Notably, GFW has responded to user needs in this area and made alert coordinates available for download/export in these ways.

There's No Internet in the Forest

This is perhaps an obvious problem, but it's a fundamentally frustrating one for GFW, which is at its essence a web-based platform. In fact, we experienced Internet connection problems during all of our interviews, and our participants often experienced problems with the GFW platform lagging or freezing across poor Internet connections during the protocols, to the point where in some cases we had to abandon the protocols and move on to the audio interview; and these interviews weren't taking place in remote forests: they were all in offices in major cities. Internet lags further resulted in over-and-under-zooming during the protocols, which increased the difficulty with articulating global and local views. All interviewees talked at length about the difficulties of trying to transport GFW data and images out into FOCs. The Peruvian site pointed out that the problem ran both ways as well: 'We can't integrate fieldwork with the platform because most of our monitors don't have Internet, unfortunately. We can use WhatsApp, but they can't send complex data this way.'

To work around the lack of Internet, most sites adopted one of the two solutions mentioned earlier: reducing visualizations to GPS coordinates that could then be loaded into an offline digital mapping app and ground-truthed from there or printing out physical maps and carrying them into the forest. What's particularly interesting about these solutions from our perspective is that they discard top-down or broadcast solutions in favor of rhizomatic, point-to-point solutions—just as trees in a forest connect and communicate with each other via root systems hidden below the visible canopy.

The Canopy Problem

Another perhaps obvious articulation problem in using GFW is the forest canopy: satellites can't see through it and often fail to differentiate illegal plantations from old-growth forest. Furthermore, they can't see anything beneath the canopy, which means that GFW inadvertently removes all non-tree species from the scene of the forest: people, animals, other plants, fungi, and insects—not to mention rocks, soil, minerals, and small bodies of water like streams and ponds.

This subtraction brings up a fundamental definitional problem that we've discussed in earlier chapters and that came up again in our interviews: what counts as a tree? Or a forest? And who gets to say? Gonzalez commented at some length on this problem from the developer standpoint:

> Every country has a different definition of forest . . . Some people may say no, no, we actually don't consider forests until the canopy height is more than whatever or the density or . . . so harmonizing that . . . as an actor we didn't want to bother anyone. Biologists have a different take on why tropical forests are more important than temperate forests. This was discussed every year, the case of dry forests, the case of temperate forest where they are covered by snow so sometimes you don't know. The case for if something is or is not deforestation or is a false positive or you are just . . . I mean at the end, you may just have a sparser forest and a bigger dry forest, not a deforested thing.

Our Indonesian participant was careful to point out that defining forests just by trees leaves out important nonhuman species. The endangered forest system in which he works 'is the only place on Earth, where the four key species are in crisis. That is rhinos, tigers, orangutan and elephant are still crisis in the ecosystem. So, it concerns us.' Djanteng and Webb both emphasized that for them, forests include people. Djanteng argued that for SAILD,

> what made us have a greater interest in forest monitoring is we noticed that with the protected areas, the occurrence of protected areas, communities have not been . . . receiving all their rights. Like during the creation process they are not very involved in the creation process of protected areas and even in the management process. So, the rights in most cases are somehow ignored and they don't see the community depends solely on the forest for their livelihoods.

Similarly, Webb discussed the problems she's encountered trying to visualize various human dimensions in GFW:

> But one thing that I am constantly thinking about . . . is how to do a better job of representing. Some of those socio-economic issues and just represent the people that are on the map, along with the biophysical information about forests. That's been a huge challenge, just because you can't see those things from space and all of the different dynamics are so place, space and contextually specific. So there's tons and tons of information about forest communities through household surveys and things like that, but once you try to kind of bring all of that together, and like put it on a map, it just doesn't tell the whole story, or can tell the wrong story. So like working with Global Witness they have you know a great story to tell. But also very disheartening. You know that database of where environmental and land defenders were killed . . . . We looked at trying to represent that on the map as a spatial layer, but they're really just wasn't any definitive correlation between conflict or violence and deforestation, because there's so many other things that are going on. We looked at the gender data and trying to see women's role in forest governance and, if that has changed. But again once you try to do that at a global level, it just doesn't tell the story.

GFW tried a couple of solutions to put people back in the forest: one of them was 'user stories' that users could upload and link to a pin on the map. However, curating and serving those stories took up more resources that GFW could provide, given that according to the organization's internal analytics, the stories were not being read, and the feature was not listed as a priority when the user base was surveyed. So, the user stories feature was discontinued, replaced by 'Mongabay Stories,' reports on de/reforestation projects in selected sites. Another was the 'People' category on the GFW blog, which is still active. But it was clear in our interviews with developers that the canopy problem was pretty intractable from a visualization standpoint, given our current media ecology. Webb pointed out that it's a problem even poachers are aware of and exploit, if we think of the canopy as extending to cloud cover:

> Whenever there's a cloud covering the forest, when the satellites take images all they see is clouds and you can't see the deforestation. And so in some places in the Congo Basin, I mean months could pass without a clear image . . . And illegal loggers and others knew this and so they're taking advantage of deliberately going out and doing these illegal clearing activities when it was raining, because they knew the satellites couldn't see them.

The solution GFW developed to the cloud cover issue, in collaboration with Wageningen University and Research, was the development of the radar-based (Sentinel-1) radar alerts, which can penetrate cloud cover; these alerts are available both as a stand-alone system and as part of the integrated alerts layer on GFW. But the larger problem of the invisibility of forest denizens below the canopy remains. Increasingly, one solution seems to be to bring the viewpoint down to canopy level—by using drones. Drones represent an interesting hybrid between satellite and grounded views of forests; they're not synoptic, but they're not ground up either. Our Indonesian informant reported their field teams using drones to check alerts that would take too long to reach by foot due to rivers, cliffs, etc. As mentioned earlier, the Peruvian NGO trains monitors to use drones to protect themselves from the potentially violent operators of illegal coca plantations. GFW developers are aware of the use of drones and in fact have done trainings with drones for ground-truthing of alerts. However, as of yet there's been no purpose-built integration of drone imagery or data with the Forest Watcher app. GFW reports prioritizing the development of new app features based on user feedback, so perhaps as more users integrate drones into their forest monitoring practices, this integration will develop.

Maps as Boundary Objects

Boundary objects are things that two communities understand quite differently, but that they both care a great deal about, which allows them to work together across their differences.[5]

> **Boundary object:** First defined by Star and Griesemer in 1989 as a matter of concern shared by two different communities who nonetheless define the matter quite differently—yet their shared concern for it allows them to articulate their actions with each other through it. A river is a classic example of a boundary object: anglers and biologists are communities with divergent values who understand a 'river' very differently, but they both prize it and can and will work together to protect it.

In our interviews, we discovered maps functioning as boundary objects, allowing agencies and communities with very different values and lifeways to communicate and work together. In this way, maps function as a major solution to the problem of articulating understandings of forests among diverse stakeholders. Gonzalez commented directly on this point:

> What we're learning is that people are moved by their own concerns, and they are looking for what they need to know. It's like some people are not going to see the map ... they just want to see the country ranking. But they didn't speak that geospatial language, right? ... You can't do a tool that fits everyone ... [but] I think that's one of the big strengths of GFW that it really creates a nice partnership around it.

Articulating NGOs and Indigenous Groups

As mentioned earlier, maps are the lingua franca between NGOs and Indigenous communities, whether they're viewed on a laptop slideshow or taped to a wall. Webb reported printed maps as being a primary strategy of communication across linguistic and cultural boundaries for GFW fieldwork:

> So usually, then I go and I print out like a big map with the satellite imagery and the alerts. And then ... typically there's a point of reference for folks within the community. Like ... a lot of the Communities live off of rivers and so you know [it helps in] orienting them. Like okay here's the River, this is where we are. And then it's helpful in that way for them to think about so it's not always like north or south, right? Like what does that mean if you're not familiar with those concepts?

Our Peruvian informants shared related images of NGO workers and Indigenous monitors standing at a printed-out map and pointing out a recognizable landmark together with their fingers. This is a classic example of a boundary object in operation: we don't need access to exactly what that landmark means to the other party; what matters is we were able to articulate a shared recognition and valuation of that landmark that would allow us to coordinate activities. The Peruvian site further reported that the maps were motivating to their monitors: 'All the analysis we do is basically prepared for the community so they can see how these deforestation dynamics grow or are built and they can learn the patterns. Many communities know deforestation is happening, but when they see these kinds of images, they feel motivated and enabled to intervene.' Our Indonesian informant also told us that maps were the main way his NGO communicated with Indigenous monitors.

Articulating NGOs and Government Agencies (Peru, Indonesia, and Phanuella)

GFW maps also work as articulation points between NGOs and government agencies that, outside of forest preservation, have very different goals and responsibilities. Webb reported maps were a major feature of the GFW Summit, held every 2–3 years, that brings NGOs and governmental representatives, among other user groups, together with data and technology providers to create a 'community of practice.' Our Indonesian participant reported maps being the centerpiece of the reports they create for the government:

> The visualization we collect, all the 100% confidence data, we produced in Google My Maps. We did a presentation to the Central Forest Agency in [redacted] in which we shared the SAT file sharing the information on forest loss and also the KMZ and KML formatted data so they can open it in Google Earth in Google maps or other platforms.

Our Peruvian participants reported similar efforts using GFW maps, particularly video ones showing deforestation increasing over time:

> Five or six years ago, we used the time-lapse videos to show how illegal coca plantations were being developed within the bounds of a popular and beloved national park, so that raised awareness … . We also have used the platform to win territorial rights for Indigenous communities, when we can demonstrate via superposition the overlap of current timber concessions or drug trafficking routes with traditional territories. We can use these to fight against nepotistic favors that are granted to concessionaires to the detriment of local communities and their traditional rights. We can also use the platform to advocate for the restoration of degraded ecosystems.

Articulating Government Agencies With Each Other, Industry, and Civil Society

As a government agent, Gigia Aleksidze was the best source of information on the use of GFW maps as boundary objects between government agencies and other stakeholders. For instance, he reported having used GFW maps to help articulate the efforts of various departments in his ministry, and with the integration of Soviet-era maps, he hoped to expand the use to other sectors of the government:

> In Soviet times they developed … very precise, very good land-use maps. And even now, sometimes we are using it, because we could see what … let's say where windbreaks were or where they were cultivating the fruit.

> So this is lacking right now, for when speaking about any political decisions or political change let's say land policy, it's very centralized and we need very good information, very good visualization, very good data. Then maybe we could start discussions for some policy developments. I can see it's very helpful for the future.

He also reported using GFW maps as boundary objects to negotiate timber concessions with companies that might not be aware of sensitive landscape areas:

> So we found out that there were also some other information and data needed for the forestry sector. For instance, let's say when the foresters, based on the forest management plans, when they allocate some of the harvesting areas, for instance, for timber cutting, sometimes we found that everything was all right—they used all the information that was provided in the forest management plans, including the slopes or density and other parameters. But it happened that we found out some of the harvesting areas were appointed in areas where there was a higher risk sort of natural disasters. For instance, there were some risks for landslides and then some other risks . . . . Also for the protected areas some of the projects and some of the infrastructure locations, we included in the data sets to visualize as different layers.

He also has used GFW maps to work with academics and other civil society stakeholders. In particular, there was a high-profile case in which the maps were used to stop logging around important cultural sites:

> There were some problems of moving this heavy machinery around the areas so that people would be disturbed with it. So there was a kind of people's interest which led to the stop of actual logging. But, of course, the platform was also used for visualization to see that area. To show it as a presentation and of course it helped as a kind of communication, even if there was no platform or even no internet or something. It was the people's will and what people's decisions were based on.

And finally Aleksidze has used the maps to connect the government with forestry students and professors in Georgia under the guise of getting feedback on the utility of the GFW platform. 'Generally when we had these communication meetings if we can call it so . . . yeah mainly we are showing what we were collecting why we are visualizing. We also wanted to get their feedback. In order to see how they could use it if we developed in a way they need it.'

Maps are hardly the only solution to all of the articulation problems reviewed earlier, but in our case studies they served as a common ground that various constituencies could rally around and recognize themselves and their interests in. And ultimately, as Gonzalez points out, maps are why users come to GFW in the first place.

Results of GFW User Survey

In 2021, a survey of roughly 300 GFW users worldwide, with 48 follow-up interviews, was carried out by an Australian research team in coordination with WRI leadership.[6] While we won't review all those results here, it's worth mentioning that they corroborated our findings from case studies as follows:

- **The problem of zoom:** Survey respondents, particularly community organizers and commodity producers, wanted higher-resolution views of their forests of concern (p. 7).
- **The wood wide web:** Survey respondents noted difficulties articulating layers and features of GFW and difficulties sharing data between GFW and related Google and GIS-based applications. The survey team noted, as did Webb in our interviews, the disconnect between WRI's intentions in serving users and the actual user base, particularly with regard to gender dynamics (p. 7). The survey team noted the difficulty of serving both high-resource and low-resource user populations with one platform and asked WRI to consider if they wanted to continue to do so or if they wanted to focus efforts on high-resource users.
- **The canopy problem:** Survey respondents reported high error rates with distinguishing crops like oil palms from primary forest from space (pp. 32–33).
- **Maps as boundary objects:** The GFW website, featuring maps and data analysis/reports of those maps, was rated the highest among all GFW products in terms of its political impact at all levels of governance and between public and private sectors (p. 37–44).

Conclusion

In this chapter, we reported the results of our empirical study of how GFW developers and users solved the problem of integrating global and local views of forests to support their EJ goals and political action. We conducted interviews with three GFW developers and four power users of GFW and protocols (in which users performed typical tasks in GFW while vocalizing what they were doing) with three of the users. From these data, and data from a 2021 GFW user survey, emerged four themes relating to the articulation of global and local views of forests; these themes were understandably related to our critical analysis, but some new topics emerged from participants as well. Key findings under those themes that we will discuss in the conclusion to this book include the following:

- **The problem of zoom:** Calls for the incorporation of more local data in GFW, and the protection of that data from global circulation, are related to Indigenous data sovereignty movements and the safety of Indigenous and local monitors.
- **The wood wide web:** Users of GFW in the Global South overcome problems of broadcasting views of forests—due to problems with satellite and radio-tower connections—via an ingenious series of point-to-point connections.
- **The canopy problem:** Drones are providing an interesting hybrid solution to the problems of articulating global and local views of the many non-tree beings that make up forests.
- **Maps as boundary objects:** Although they continue to import transnational neoliberalism into local and Indigenous contexts in the Global South, maps also materialize Indigenous peoples, their territories, and their concerns at geopolitical scales. Therefore, they remain the primary—and in many cases the only—*lingua franca* for negotiating forest policy among polities at all levels.

In the final chapter of this book, we will consider the implications of our GFW case study for the larger problem of articulating global and local views of forests in forest monitoring. We argue that the case study suggests as a promising way forward the concept of *storyworld networking* or connecting cosmograms of forests into networks that tell compelling stories. A storyworld network offers, as an antidote to the green marble of global panopticism, a robust and resilient assemblage of point-to-point articulations among agents in forest conservation—*much like the networks formed by forests themselves*. A storyworld network also generates concrete recommendations for both designers and users of global forest monitoring platforms like GFW.

Notes

1 "The State of the World's Forests: Forests, Biodiversity, and People," 2020, accessed August 13, 2023, www.fao.org/3/ca8642en/ca8642en.pdf.. Unless otherwise stated, all general forest facts about our study sites come from this report.
2 Brittany L. Peterson, "Thematic Analysis/Interpretive Thematic Analysis," *The International Encyclopedia of Communication Research Methods* (Hoboken: John Wiley & Sons, 2017) 1-9.https://doi.org/10.1002/9781118901731.iecrm0249.
3 Horton, *Cosmic*, 6–8.
4 Tim Cadman et al., *Evaluation of Global Forest Watch: Final Report* (Griffith University, 2021), 7.
5 Susan Leigh Star and James R. Griesemer, "Institutional Ecology, 'Translations' and Boundary Objects: Amateurs and Professionals in Berkeley's Museum of

Vertebrate Zoology, 1907–39," *Social Studies of Science* 19, no. 3 (August 1, 1989), https://doi.org/10.1177/030631289019003001, http://sss.sagepub.com/content/19/3/387.abstract.

6 Cadman et al., *Evaluation of Global Forest Watch: Final Report.*

Bibliography

Cadman, Tim, et al. *Evaluation of Global Forest Watch: Final Report.* Griffith University, 2021.

"Cartographies of the Unseen." 2021, https://research.felipecastelblanco.com.

Horton, Zachary. *The Cosmic Zoom: Scale, Knowledge, and Mediation.* Chicago, IL: University of Chicago Press, 2021.

Peterson, Brittany L. "Thematic Analysis/Interpretive Thematic Analysis." *The International Encyclopedia of Communication Research Methods* (2017): 1–9.

Star, Susan Leigh, and James R. Griesemer. "Institutional Ecology, 'Translations' and Boundary Objects: Amateurs and Professionals in Berkeley's Museum of Vertebrate Zoology, 1907–39." *Social Studies of Science* 19, no. 3 (August 1, 1989): 387–420, https://doi.org/10.1177/030631289019003001, http://sss.sagepub.com/content/19/3/387.abstract.

"The State of the World's Forests: Forests, Biodiversity, and People." 2020, accessed August 13, 2023, www.fao.org/3/ca8642en/ca8642en.pdf.

7 From Green Marbles to Storyworlds

Insights From Critical Analysis of Global Forest Visualization Platforms

In Chapters Two and Three, we traced the history of global forest visualization through four epochs that roughly correspond to Turner's spheres, each with its own signature cosmogram:

- **Commodification:** Starting in the late middle ages in Europe, perhaps earlier in Imperial China and Africa, forests began to be owned and inventoried as wealth. The cadastral (survey) map is the signature cosmogram of this epoch.
- **Capitalization:** In the Anthropocene, forests were gridded, managed as plantations, and categorized for profit in a global colonial-capitalist trade regime. The geodesic projection map is the signature cosmogram of this epoch with its grid of longitudinal and latitudinal lines.
- **Globalization:** With the advent of satellite imagery of the Earth and a growing awareness of climate change, forests became thought of as global, the 'lung' of the planet. The green marble is the signature cosmogram of this epoch.
- **Glocalization:** Interactive, Internet-based mapping platforms utilizing satellite imagery make it possible to 'zoom' among scales ranging from the global to 30 m^2 in order to view forests at multiple scales; these developments go hand-in-hand with increasing efforts at transnational levels (e.g., the UN) to monitor and manage forests, particularly tropical forests, to mitigate climate change. The signature cosmogram of this epoch is the Google map.

This history led us to two key insights about the geopolitics of zoom-based visualization platforms:

1. They make a default visual argument that whoever achieves the power to see all the world's forests at once *synoptically* also has the power and

DOI: 10.4324/9781003376774-7

responsibility to manage the world's forests for the good of their inhabitants. This default visual argument has substantially participated in a politics in which well-resourced nations in the Global North have been licensed to dictate, to various degrees, forest policies in the Global South (the current focus of global forest conservation) through programs like Reducing Emissions from Deforestation and Forest Degradation (REDD+).
2. Zoom-based visualizations—because of issues with server capacity and data standards—have a tendency to ignore local data that conflict with global data or reconcile them to the global standard of measurement.

We named this particular brand of geopolitics *Google Gaia*. This concept builds off the Gaian concept of the Earth as a being, conceived by Lovelock and Margulis and developed by Latour and others; also, Gurevitch's 'Google Warming' is the conviction that we can control the world from a keyboard that arises from the conflation of the real world with the computerized model of it. Accordingly, Google Gaia is a planetary being whose wellbeing we can steward through a computer terminal.

Google Gaia is one of many kinds of *cosmograms*, a term we borrowed from John Tresch to indicate a community's image of the world that communicates not only what the world is like but also how we should live in it, that is, its *cosmology*. As discussed at length in Chapter Four, a cosmogram is not just a picture, it's a piece of propaganda, a framing narrative for communal life, a political morality tale. And due to its media history and ecology, Google Gaia is a cosmogram that participates in *cosmological imperialism*, a kind of geopolitics that seeks to impose the cosmology of the neo/colonial nation state (or set of nation states) on colonized communities, either in part or whole. In other words, Google Gaians think they can fix the planet because in their models forest-ecological problems always work out if the parameters are set correctly. Plantations and geo-engineering are the logical solutions to a problem seen only from space. So the logic of the Google Gaia Engine is yet another justification for a green marble, instrumental stewardship of the Earth because the only ones who can see the whole problem are the only ones licensed to fix the whole problem.

From an environmental justice (EJ) perspective, cosmological imperialism is obviously problematic, and so alternative cosmologies and cosmograms have risen up to counter it. We reviewed two major types of alternative cosmograms in Chapter Four: counter-maps and dwellings. Both enact a *spherical* or bottom-up/inside-out cosmology to oppose *global* (top-down/outside-in) synoptic cosmograms of forests. Counter-maps disrupt synoptic maps by heavily layering them with coded information only local communities can read or by creating 'soundscapes' and other maps that can only be used on the ground or through other senses than the visual. Dwellings create encompassing spaces or bubbles that have to be inhabited to be understandable; they're opaque and illegible to outsiders.

Some EJ activists have chosen counter-maps and dwellings to achieve their goals. But others still find value in synoptic global forest visualization tools. We set ourselves the task, then, of discovering strategies those EJ activists can use to mitigate or counteract the cosmological imperialism of global forest visualizations. As a rubric for evaluating these strategies, we developed the critical framework of *storyworld networking*. We got *storyworld* from a literal translation of cosmology and *networking* from critical work, which aims to articulate global and local worldviews without reducing one to the other. With this framework in hand, we turned in Chapters Five and Six to our case study of GFW, an open-source platform with explicit EJ goals, to see how well it was able to meet our criteria. We also wanted to discover creative solutions that GFW's users have invented to meet their own goals despite the limitations of the platform.

Results of the Case Study of GFW

The Desktop Analysis

We began by applying our storyworld networking criteria to a desktop analysis of GFW: in other words, we looked at the platform to see to what extent it allowed the articulation of local and global views of forests without reducing one to the other. As we are both critical Humanists, we took into account not only the visible and interactive features of the platform but also its mission, its media ecology (how it supports and is supported by other media and infrastructure), its history, and its geopolitics (who uses it and why, who invests in it and why, who gets to define what counts as a forest in it, and why).

We concluded that, based on its history and media ecology, GFW does count as a Google Gaia cosmogram. As such, by default and in many cases against the express wishes of its designers, it participates in cosmological imperialism, the traces of which are apparent in its dominant user base (white, educated men from the Global North), its investors both former and current (including large transnational palm oil traders like Cargill and Unilever), its use by national government agencies, the structural difficulties with uploading and serving local forest data in the global platform, and the present emphasis on developing GFWPro, which targets transnational, commercial traders and investors in forest products.

However, a Google Gaia cosmology is not the only one that can be built from the rich and diverse media ecology of GFW. We found encouraging evidence of a storyworld network operating in the multiple geopolitical layers that users can load, including climate change and biodiversity; in the Map-Builder desktop client that allows users to integrate local data and share them securely with local stakeholders; in the pinned stories from Mongabay and the blog features accompanying the maps; and in the Forest Watcher mobile app and its ability to geo-tag photos of deforestation taken on the ground. The storyworld network created via these features links global and spherical images of the Earth into a complex cosmogram that can support EJ action. At this point in the analysis, we needed to investigate how users put these features into action in order to be able to more fully evaluate GFW as a potential storyworld network.

Results of User Protocol Interviews

As detailed in Chapter Six, we conducted interviews with three developers/managers of GFW and four users; with three of these users, we also conducted think-aloud protocols, which permitted us to observe the users in action with the platform. Findings emerged along four themes having to do with problems and solutions articulating global and local views in GFW:

- **The problem of zoom:** Users and developers confirmed difficulties in incorporating local data in the global platform, and in validating the global views locally, but all users reported work-arounds such as choosing particular landmarks as reference points around which to compare views at various scales; ground-truthing global data on foot with monitors and drones; and exporting GFW data into Google Earth and other platforms that more easily integrated geospatial data from various data standards.
- **The wood wide web:** Problems with infrastructure, technological literacy, and Internet access were frustrating for all users, but they had developed ingenious solutions such as exporting data points to send as small files across cellular networks; printing out and hand-carrying maps into forests for monitors to use; and empowering local Indigenous community leaders (often women) to learn and teach technical skills to monitors.
- **The canopy problem:** Both developers and users struggled to integrate views above and below the forest canopy into a complete picture of the forest ecology, but once again, these views were able to be articulated through ground-truthing, the use of drones that operate at canopy height, and the layering of maps containing vital information about non-tree forest species including animals, plants, and humans.
- **Maps as boundary objects:** It was clear from our interviews that no two communities (perhaps even no two people) understand forests and forest maps in precisely the same ways. However, it is exactly this misunderstanding and slippage in articulation that leads to rich and productive exchanges between stakeholder groups that care about forests but for very different reasons. We observed forest maps operating as boundary objects and facilitating cooperation among government agencies, civil society, nonprofit organizations, Indigenous communities, academic communities, and industry partners.

General Conclusions

Perhaps the most useful general conclusion we can draw from our research into global forest visualization is that the more we can learn how to see forests *from* the forests themselves, the more equitable and actionable those visualizations will be.

What do we mean by learning to see forests *from* forests? We mean this: we speak of both a tree and a forest as coherent beings. Yet each forest is a network of trees, and each tree is a network of systems (leaves, woody tissue, vascular tissue, and roots); those systems are networks of even smaller cellular systems; and those systems are networks of molecular systems. At the macro level, forests can (occasionally and in the right conditions) network with other forests to communicate chemical information about drought, predation, and disease.[1] This fractal structure of forests provides us with a model for articulating multiple visualizations of forests—even if they overlap and disagree—into a coherent view that tells a story we can act upon. And indeed this is what we saw happening in our developer and user interviews, with all the ingenious articulations and work-arounds that GFW users developed to overcome difficulties in reconciling global and local views of forests.

Of course, in imagining a storyworld network as a productive cosmology for forest visualization, we are building on a substantial body of previous theory. Chief among this work is the 'rhizomatic' theory of articulation traced in Deleuze and Guattari's *Thousand Plateaus*. Put forward as an explicit antidote to top-down, causal reductionism (in which things are reduced to their causes, as in 'communism made this happen' or 'that's because she's a woman'), Deleuze and Guattari insist on connecting the events and agents of history side-by-side as nodes in a network, linked according to their interactions and refusing to let one node swallow the others. 'It is odd how the tree has dominated Western reality,' they write, and they propose instead that we look below the surface at the rhizomatic connections in the soil, which generate 'a map that is always detachable, connectable, reversible, modifiable, and has multiple entryways and exits and its own lines of flight.'[2]

Although developed independently, Latour's actor-network theory offers a related logical foundation for the analysis of climate change and other messy natural-cultural phenomena. Specifically for visualization, Latour and his followers recommend 'terrestrial' visions of the Earth instead of global ones, which articulate many layers and perspectives into a coherent, but not homogenous, view of the 'few kilometers' of the planet's surface that support all known life: 'Gaïa-graphy' of the Critical Zones and the principle of the Anti-Zoom, for example, have been suggested as ways of generating maps that cannot be 'landed on' but must rather be lived in.[3]

Arturo Escobar describes a similar solution when he posits a 'pluriverse' as a comprehensive environmental vision—one composed not of a single god's-eye view of the environment but rather of a network of potentially infinite, often-overlapping views.[4] New theories of design arise in a pluriversal context, and Escobar likes in this instance the 'transition design' program at Carnegie-Mellon, whose principles sound very much like our storyworld networking criteria, particularly 'Uses living system theory as an approach to understanding/addressing wicked problems,' 'advocates place-based, globally networked solutions,' and 'links existing solutions so that they become steps in a larger transition vision.'[5]

Donna Haraway suggests a related solution in her 'carrier bags' post-script to *Critical Zones*. In the concept of the 'carrier-bag,' which is a woven network, she posits a framework that 'proposes and tests worlds so as to render readers more attuned to difference, to possibilities, to other ways of living and dying not trapped in the endless cyclopean war story from above.'[6] And while not strictly a visualization theory, Richard Powers's *Overstory* of course provides a concrete example of how multiple storyworlds can be linked into a coherent 'overstory'; furthermore, it's an example founded explicitly on the structure of trees and forests themselves, where many branches meet in a coherent tree, and multiple diverse species come together in a forest.[7]

Examples of Storyworld Networks in Forest and Environmental Studies

What do some concrete examples of storyworld networks for forest conservation look like? We have prepared a convenient handlist of examples as an appendix, with links, but here we will discuss a few that we have found the most inspirational.

We have already mentioned multiple times the book *Critical Zones: Observatories for Earthly Politics*. It and the companion web exhibition from ZKM Karlsruhe are a treasure trove of innovative visualizations of forests and other terrestrial systems. For instance, the 'Atmospheric Forest' observation by Smite and Smits combines a 360-degree video of a forest suffering from drought in Switzerland with visualizations of volatile compound loss and hydrophone audio of the trees crackling like fire as they dry out.[8]

Forensic Architecture is a network of multidisciplinary investigators who work to reconstruct scenes of crimes against human rights. Their multimodal arguments about the devastating effects of mining on Brazilian rainforests are an excellent example of layering cell phone footage of environmental crimes over global maps of deforestation, very much like GFW maps, and linking these into a compelling video story about the geopolitics of forest degradation (see Figure 7.1 for a screen capture of the video).[9]

The Environmental Justice Atlas is a vital resource for Environmental Justice (EJ) work and also contains several storyworld networks around forest conservation. An excellent example is a 'featured map' on the social problems following India's designation of forest preserves as protected areas. The map is accompanied by a blog-style story that makes the argument that the preserves

Figure 7.1 Still from 'Gold Mining and Violence in the Amazon Rain Forest.'

Source: Figure adapted from https://forensic-architecture.org/investigation/gold-mining-and-violence-in-the-amazon-rainforest.

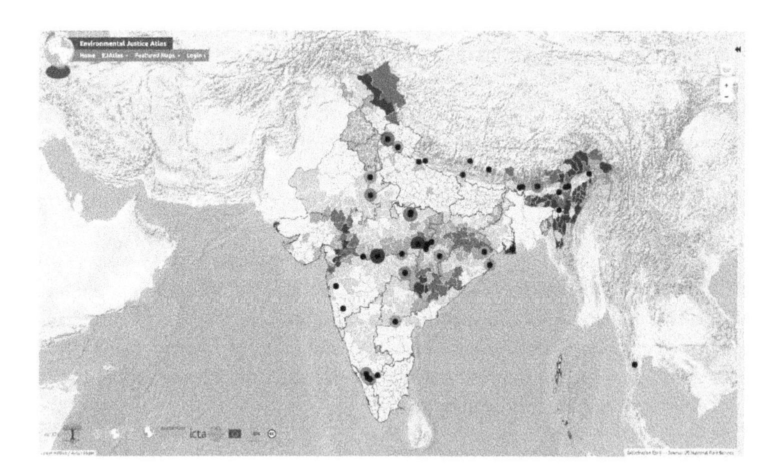

Figure 7.2 Screen capture from 'Losing ground: How are India's conservation efforts putting the local communities in peril?,' feature map by Fanari et al., *Environmental Justice Atlas, 2019*

Source: Figure adapted from https://ejatlas.org/featured/conflictprotectedareaindia.

are intentionally and unintentionally exiling minority and Indigenous groups that have had conflicts with the government over traditional land rights: map layers of conflicts, preserve areas, and Indigenous groups, when loaded together, support the article's argument by creating darker and darker regions in problematic areas as they intersect (Figure 7.2); in addition, stories from multiple sources about the problem are pinned to the map in the relevant areas.[10]

Finally, *Cartographies of the Unseen* is a collaborative 'storymap' curated by artist Felipe Castelblanco that seeks to visualize as many stories as possible linked by the 'vertical axis' of contested sites in the Colombian Pan-Amazonian rainforest. Castelblanco explains the storymap concept as follows: 'by using sensorial ethnography, forms of counter-mapping, participatory art, writing, video and installations, I seek to address planetary entanglements across different layers of space, from the underground to exosphere and in apparently remote and disconnected locations from the Global South.'[11]

All of these examples meet our storyworld networking criteria in that they interpolate both global and local views of forests, without reducing one to the other, to create coherent visual arguments for specific EJ actions. They serve as holistic inspirations. In what follows, we will offer a list of practical suggestions for both designers and users of global forest visualization platforms based on our research.

Recommendations for Designers/Producers and Users of Global Forest Visualization Platforms

Designers and producers of Global Forest Visualization Platforms should

- build in safeguards for client data privacy and make data sharing settings transparent and flexible;
- consider integrating drone GPS and camera data with handheld/field forest mapping apps; focus development on mobile applications;
- make it easy to print and export maps, videos, and GIS data in multiple formats (including a dedicated grayscale mode that boosts contrast on boundaries, landmarks, and place names for black-and-white printing);
- at least in desktop applications, facilitate importation and layering of as many GIS data standards as possible and build in an option to pin photos, videos, and audio files to specific locations; and
- produce written briefs/reports on topics and areas of interest, as 'Written reports on topics of interest' was the top-ranked information format in the GFW user survey, followed by maps.[12]

Users of global visualization platforms should

- be aware that by default, if not actively counteracted with other views, satellite views of forests encourage transnational, neoliberal solutions to forest management;

- always be aware of privacy settings in apps and don't use them if you don't know who's seeing your data;
- never give away data about your territory, community, or culture without (a) knowing where they're going and how they're going to be used and (b) getting equal value in return, in terms of something that your community needs;
- remember that maps and data always tell stories—and by default they will tell the story of the dominant group in power, so be aware of that and make sure you tell your story when you present a map or a dataset;
- use open-source multimedia software like kdenlive, PiTiVi, or Astrofox to combine as many different views of your forest of concern as possible;
- use forest monitoring as an opportunity to teach and empower women and other community members with technical literacy; at the same time, be aware that that literacy infects anyone who has it to some degree with the values of the culture who make and sells the technology; and
- as far as possible, form connections with other forest communities who share similar concerns—share information and resources.

Conclusion

We began this book with a story—a story about a Nicaraguan farmer coming out to his field to find a European analyst there performing an assessment of the health of the trees growing there according to criteria developed by Canadian foresters, an assessment which would result in the farmer not being paid what he had anticipated in the REDD+ program he had torn out food crops to participate in. This was a true story, recounted to us by a friend who had done fieldwork in that area of Nicaragua, and while it was not necessarily representative of all REDD+ projects, it nonetheless encapsulated some bigger dynamics that intrigued and troubled us as researchers in visual rhetoric and media ecology: namely, why did the European analyst feel he had the right to be in a field that wasn't his making an assessment that would affect the livelihood of the field's owner? And how had he gotten there in the first place even though he was unable to communicate with the locals? The answers to those questions are of course complex, but we were interested in the parts that related to visualization: the view of the forest the analyst had worked with on his computer in Europe and the GPS map he had used to navigate his rental car to that Nicaraguan field.

That story led us to GFW, whose mission was explicitly pro-EJ and which, with its open-source platform and support for non-profit and Indigenous community organizations, was seeking to counteract those panoptic neoliberal dynamics that had troubled us so much about the Nicaraguan story. The GFW team welcomed our inquiries and read our papers, even when we had critical notes about the platform they had worked for two-plus decades to build and grow. They had multiple meetings with us, freely shared reports from their user survey, and connected us to developers and power users who helped us to deeply investigate the questions we wanted to ask about global forest visualization: what counts as a tree, or a forest, from space? Who gets to say? How do we get from space to below the canopy to see all the non-tree species that make up the forest? How can global and local views of forests be combined without flattening one to the other?

Answering those questions turned into another story, which you have just read—the story of how we got interested in forest visualization, how those interests evolved and narrowed into this project, how the developers and users of GFW continue to work for EJ in and around the constraints imposed by the Google Gaia cosmology, and how we concluded that the more we learn to see forests from forests themselves, the better we will see them. We end with the sincere hope that our concept of storyworld networking, the examples we have curated, and the recommendations we have made will help the generous people who helped us create more just and equitable ways of seeing the forests on whose lives ours quite literally depend.

Notes

1 G. Arimura and I. S. Pearse, "Chapter One—From the Lab Bench to the Forest: Ecology and Defence Mechanisms of Volatile-Mediated 'Talking Trees'," in *Advances in Botanical Research*, ed. Guillaume Becard (London: Academic Press, 2017) 3-17; Teja Tscharntke et al., "Herbivory, Induced Resistance, and Interplant Signal Transfer in Alnus Glutinosa," *Biochemical Systematics and Ecology* 29, no. 10 (November 1, 2001), https://doi.org/https://doi.org/10.1016/S0305-1978(01)00048-5.

2 Gilles Deleuze and Félix Guattari, *A Thousand Plateaus: Capitalism and Schizophrenia* (Bloomsbury Publishing, 1988), 18–21.

3 Alexandra Arènes, Bruno Latour, and Jérôme Gaillardet, "Giving Depth to the Surface: An Exercise in the Gaia-Graphy of Critical Zones," *The Anthropocene Review* 5, no. 2 (2018), https://doi.org/10.1177/2053019618782257; Bruno Latour, "Anti-Zoom," in *Contact, catalogue de l'exposition d'Olafur Eliasson* (Paris: Fondation Vuitton, 2014).

4 Arturo Escobar, *Designs for the Pluriverse: Radical Interdependence, Autonomy, and the Making of Worlds* (Durham, NC: Duke University Press, 2018), 67.

5 Ibid., 158.

6 Donna Haraway, "Carrier Bags for Critical Zones," in *Critical Zones: The Science and Politics of Landing on Earth*, ed. Bruno Latour and Peter Weibel (Cambridge, MA: MIT Press, 2020), 440.

7 Richard Powers, *The Overstory: A Novel* (New York: WW Norton & Company, 2018).

8 Bruno Latour and Peter Weibel, *Critical Zones: Observatories for Earthly Politics* (Cambridge, MA: MIT Press, 2020); "Atmospheric Forest, 2019-Ongoing," *ZKM Karlsruhe*, 2020, accessed August 14, 2023, https://critical-zones.zkm.de/#!/detail:atmospheric-forest.

9 "Gold Mining and Violence in the Amazon Rain Forest," *Forensic Architecture*, 2022, accessed July 20, 2023, https://forensic-architecture.org/investigation/gold-mining-and-violence-in-the-amazon-rainforest.

10 "Losing Ground: How Are India's Conservation Efforts Putting the Local Communities in Peril?" *Environmental Justice Atlas*, 2019, accessed July 30, 2023, https://ejatlas.org/featured/conflictprotectedareaindia.
11 Felipe Castelblanco, "Cartographies of the Unseen," 2021, https://research.felipecastelblanco.com.
12 Tim Cadman et al., *Evaluation of Global Forest Watch: Final Report* (Griffith University, 2021), 30.

Bibliography

Arènes, Alexandra, Bruno Latour, and Jérôme Gaillardet. "Giving Depth to the Surface: An Exercise in the Gaia-Graphy of Critical Zones." *The Anthropocene Review* 5, no. 2 (2018): 120–35, https://doi.org/10.1177/2053019618782257.

Arimura, G., and I. S. Pearse. "Chapter One—from the Lab Bench to the Forest: Ecology and Defence Mechanisms of Volatile-Mediated 'Talking Trees'." In *Advances in Botanical Research*, edited by Guillaume Becard, 3–17. London: Academic Press, 2017.

"Atmospheric Forest, 2019-Ongoing." *ZKM Karlsruhe*, 2020, accessed August 14, 2023, https://critical-zones.zkm.de/#!/detail:atmospheric-forest.

Cadman, Tim, et al. *Evaluation of Global Forest Watch: Final Report*. Griffith University, 2021.

Castelblanco, Felipe. "Cartographies of the Unseen." 2021, https://research.felipecastelblanco.com.

Deleuze, Gilles, and Félix Guattari. *A Thousand Plateaus: Capitalism and Schizophrenia*. London: Bloomsbury Publishing, 1988.

Escobar, Arturo. *Designs for the Pluriverse: Radical Interdependence, Autonomy, and the Making of Worlds*. Durham, NC: Duke University Press, 2018.

"Gold Mining and Violence in the Amazon Rain Forest." *Forensic Architecture*, 2022, accessed July 20, 2023, https://forensic-architecture.org/investigation/gold-mining-and-violence-in-the-amazon-rainforest.

Haraway, Donna. "Carrier Bags for Critical Zones." In *Critical Zones: The Science and Politics of Landing on Earth*, edited by Bruno Latour and Peter Weibel, 440–45. Cambridge, MA: MIT Press, 2020.

Latour, Bruno, and Peter Weibel. *Critical Zones: Observatories for Earthly Politics*. Cambridge, MA: MIT Press, 2020.

"Losing Ground: How Are India's Conservation Efforts Putting the Local Communities in Peril?" *Environmental Justice Atlas*, 2019, accessed July 30, 2023, https://ejatlas.org/featured/conflictprotectedareaindia.

Powers, Richard. *The Overstory: A Novel*. New York: WW Norton & Company, 2018.

Tscharntke, Teja, Sabine Thiessen, Rainer Dolch, and Wilhelm Boland. "Herbivory, Induced Resistance, and Interplant Signal Transfer in Alnus Glutinosa." *Biochemical Systematics and Ecology* 29, no. 10 (November 1, 2001): 1025–47, https://doi.org/10.1016/S0305-1978(01)00048-5.

Appendix

Storyworld Network Resources

A handlist of online examples of visualizations of forest and climate that substantially succeed in networking local and global images into a non-reductive, coherent story that can motivate environmental justice action.

Visualizations of Climate and Environment in General

The Feral Atlas: As described in Chapters Four and Seven, this project by a consortium of environmental researchers, designers, and artists re-maps environmental crises to take into account their Indigenous histories and geopolitical causes: https://feralatlas.org/

Critical Zones: This companion project to the book by Weibel et al. presents multimedia counter-maps of a variety of environmental systems currently in crisis: https://critical-zones.zkm.de/#!/

EJ Atlas: A collaborative investigative project coordinated by researchers at the University of Barcelona, it seeks to map the intersection of environmental and social crises in multi-layered ways: https://ejatlas.org/

Forensic Architecture: While focused specifically on abuses of human rights, the multimedia projects undertaken by this multidisciplinary collective frequently have environmental intersections: https://forensic-architecture.org/

Ciclos Tiquié: Discussed in Chapter Four, this project shares Indigenous cosmologies of the Rio Tiquié in a multi-layered cosmographic calendar: https://ciclostiquie.socioambiental.org/en/index.html

Transition Design at CMU: Mentioned in Chapter Seven, this decolonial design program at Carnegie-Mellon shares many principles with the storyworld networking criteria we developed in our project: https://design.cmu.edu/tags/transition-design

Bureau d'études 'Champs Electromagnetique': French artists who produce 'cartographies of contemporary political, social and economic systems,' here they engage a project mapping electrical infrastructure: https://bureaudetudes.org/category/champ-electromagnetique/

Simone Fehlinger's 'New Weather TV': A designer, videographer, and researcher at the Deep Design Lab at the Cité du design-ESADSE in Saint-Etienne (France), she explores fiction-based realities by questioning the performativity of design and its ability to create ideologies through form. Here, she imagines a new way of forecasting weather that is postcolonial and pluriversal: www.citedudesign.com/fr/a/teasing-new-weather-tv-post-producing-global-views-1092

Visualizations of Forests

Mapa Oaxaca: This project layers traditional Oaxacan Indigenous boundaries over GFW maps so that ejidarios or local custodians can better monitor illegal deforestation in their territories: www.globalforestwatch.org/blog/people/mapa-oaxaca-platform-puts-the-power-of-data-into-the-hands-of-mexican-communities/

Atmospheric Forest: As discussed in Chapter Seven, this project from Critical Zones incorporates data from multiple sensors in a Swiss forest suffering from drought into a multi-sensory map for viewers: https://critical-zones.zkm.de/#!/detail:atmospheric-forest

Conflicts around Protected Forest Areas in India: Discussed in Chapter Seven, this EJ Atlas project uses multiple visual layers and stories to document intersections in human and environmental conflicts around protected areas in India: https://ejatlas.org/featured/conflictprotectedareaindia

Cartographies of the Unseen: This collaborative project coordinated by artist Felipe Castelblanco, and reviewed in Chapter Seven, aims to visualize Indigenous knowledge and rights in traditional territories in the Brazilian Amazon: https://research.felipecastelblanco.com

Violence and Mining in Brazilian: Mentioned in Chapter Seven as well, this Forensic Architecture project documents the causal link between illegal mining activities in the Brazilian Amazon and violence against the Indigenous peoples who live there: https://forensic-architecture.org/investigation/gold-mining-and-violence-in-the-amazon-rainforest

The Art of Hélio Melo: As described in Chapter Four, Melo was a self-taught artist and professional rubber tapper who created story-maps of the Brazilian rainforest: https://universes.art/en/magazine/articles/2008/helio-melo

Index

For Product Safety Concerns and Information please contact our EU
representative GPSR@taylorandfrancis.com
Taylor & Francis Verlag GmbH, Kaufingerstraße 24, 80331 München, Germany

www.ingramcontent.com/pod-product-compliance
Ingram Content Group UK Ltd.
Pitfield, Milton Keynes, MK11 3LW, UK
UKHW021055080625
459435UK00003B/13

9781032454009